Color Trends and Selection for Product Design

PLASTICS DESIGN LIBRARY (PDL)
PDL HANDBOOK SERIES
Series Editor: Sina Ebnesajjad, PhD (sina@FluoroConsultants.com)
President, FluoroConsultants Group, LLC
Chadds Ford, PA, USA
www.FluoroConsultants.com

The **PDL Handbook Series** is aimed at a wide range of engineers and other professionals working in the plastics industry, and related sectors using plastics and adhesives.

PDL is a series of data books, reference works and practical guides covering plastics engineering, applications, processing, and manufacturing, and applied aspects of polymer science, elastomers and adhesives.

Recent titles in the series

Biopolymers: Processing and Products, Michael Niaounakis (ISBN: 9780323266987)
Biopolymers: Reuse, Recycling, and Disposal, Michael Niaounakis (ISBN: 9781455731459)
Carbon Nanotube Reinforced Composites, Marcio Loos (ISBN: 9781455731954)
Extrusion, 2e, John Wagner & Eldridge Mount (ISBN: 9781437734812)
Fluoroplastics, Volume 1, 2e, Sina Ebnesajjad (ISBN: 9781455731992)
Handbook of Biopolymers and Biodegradable Plastics, Sina Ebnesajjad (ISBN: 9781455728343)
Handbook of Molded Part Shrinkage and Warpage, Jerry Fischer (ISBN: 9781455725977)
Handbook of Polymer Applications in Medicine and Medical Devices, Kayvon Modjarrad & Sina Ebnesajjad (ISBN: 9780323228053)
Handbook of Thermoplastic Elastomers, Jiri G Drobny (ISBN: 9780323221368)
Handbook of Thermoset Plastics, 2e, Hanna Dodiuk & Sidney Goodman (ISBN: 9781455731077)
High Performance Polymers, 2e, Johannes Karl Fink (ISBN: 9780323312226)
Introduction to Fluoropolymers, Sina Ebnesajjad (ISBN: 9781455774425)
Ionizing Radiation and Polymers, Jiri G Drobny (ISBN: 9781455778812)
Manufacturing Flexible Packaging, Thomas Dunn (ISBN: 9780323264365)
Plastic Films in Food Packaging, Sina Ebnesajjad (ISBN: 9781455731121)
Plastics in Medical Devices, 2e, Vinny Sastri (ISBN: 9781455732012)
Polylactic Acid, Rahmat et. al. (ISBN: 9781437744590)
Polyvinyl Fluoride, Sina Ebnesajjad (ISBN: 9781455778850)
Reactive Polymers, 2e, Johannes Karl Fink (ISBN: 9781455731497)
The Effect of Creep and Other Time Related Factors on Plastics and Elastomers, 3e, Laurence McKeen (ISBN: 9780323353137)
The Effect of Long Term Thermal Exposure on Plastics and Elastomers, Laurence McKeen (ISBN: 9780323221085)
The Effect of Sterilization on Plastics and Elastomers, 3e, Laurence McKeen (ISBN: 9781455725984)
The Effect of Temperature and Other Factors on Plastics and Elastomers, 3e, Laurence McKeen (ISBN: 9780323310161)
The Effect of UV Light and Weather on Plastics and Elastomers, 3e, Laurence McKeen (ISBN: 9781455728510)
Thermoforming of Single and Multilayer Laminates, Ali Ashter (ISBN: 9781455731725)
Thermoplastics and Thermoplastic Composites, 2e, Michel Biron (ISBN: 9781455778980)
Thermosets and Composites, 2e, Michel Biron (ISBN: 9781455731244)

To submit a new book proposal for the series, or place an order, please contact David Jackson, Acquisitions Editor
david.jackson@elsevier.com

Color Trends and Selection for Product Design

Every Color Sells A Story

Doreen Becker

ELSEVIER

AMSTERDAM • BOSTON • HEIDELBERG • LONDON
NEW YORK • OXFORD • PARIS • SAN DIEGO
SAN FRANCISCO • SINGAPORE • SYDNEY • TOKYO
William Andrew is an imprint of Elsevier

William Andrew is an imprint of Elsevier
The Boulevard, Langford Lane, Kidlington, Oxford, OX5 1GB, UK
50 Hampshire Street, 5th Floor, Cambridge, MA 02139, USA

British Library Cataloguing-in-Publication Data
A catalogue record for this book is available from the British Library.

Library of Congress Cataloging-in-Publication Data
A catalog record for this book is available from the Library of Congress.

ISBN: 978-0-323-39395-9

For Information on all William Andrew publications
visit our website at http://www.elsevier.com/

Publisher: Matthew Deans
Acquisition Editor: David Jackson
Editorial Project Manager: Jennifer Pierce
Production Project Manager: Jason Mitchell
Designer: Mark Rogers

Dedication

This book is dedicated to those who struggle daily with colors. This includes my technical colleagues in the laboratory, the design teams who envision inspired products, the chemists and engineers who characterize and perfect these colors, all of my friends from the Color Marketing Group who spend their weekends and free-time pouring over color boards and debating the need for a new shade of green or eliminating them entirely from a certain demographic in a specified time period.

Contents

Section IV: Color Selection Process

Acknowledgments

I have tapped into many public, invaluable resources including The Smithsonian Museum of American History, The Metropolitan Museum of Art, The Copper-Hewitt Design Museum, The Deutsches Museum in Munich, The Philadelphia Museum of Art, Museo del Novecento in Milan, The Jewish Museum in Warsaw, The Brooklyn Museum in New York City and the Museum of Fine Arts in Boston.

I would also like to express my gratitude to my fellow board members of the Society of Plastics Engineers—Color & Appearance Division who have tolerated all of my "trendy" talks over the years and taught me everything I could imagine about color in plastics, especially Dr Steven Goldstein who passed away while I was writing this book; a wonderful mentor and a true friend. I would also like to thank my colleagues at A. Schulman, Inc. who have permitted me to put this book together, specifically my colleagues in the Custom Performance Colors Group. This book would not have been possible without many years of color trending information from Trend Lab NY and all of their guidance. A debt of gratitude is extended to Lee Zevy who illuminated the importance of written expression and how it is as essential to my nature as breathing.

I would also like to thank my family: MJ and Grant who tolerate my long absences while I travel and write. Their patience and support is immeasurable and endearing.

Introduction

Colors can be comforting, provocative, seductive or even offensive. Frequently colors for products are selected by personal or committee preferences rather than what the market requires. It is also frequently one of the last decisions made in the design process. The impact of the color in a product launch can mean the difference between a raging success or a complete failure. Every new product that comes into the market has a story behind it and that story usually consists of why the product was developed, for whom the product was developed, and also the story of the developer and designer as well as their motivations for creating this product. Since we live in such an image-driven society, one of our first impressions of a product is visual. Internet advertising and print media only allow for a two-dimensional snap-shot of a product's shape and relative size. The third dimension that is instantly perceived is color. Most innovative products are developed based on unmet needs, solutions to problems or an opportunity to improve life experiences of the consumer. Generally, a great deal of time and attention is paid to the functionality and style of the product to create a product of desire. Choosing the correct color or pallet of colors will sell the story of these products and their intended path for commercial success. There is a science, history and artistic perception behind color selection and its perceived value. It is frequently over-looked and often elusive. It is the intention of this book to elucidate this process and provide guidance to the industries that need color to sell the story of their products.

SECTION I

Color Trends: An Historical Perspective

Introduction

This section will outline dominant colors and trends of each era. This is a distillation of these time periods and their significant colors. One can easily identify colors or important events that were evident in these eras that are not mentioned. The intention is to highlight colors and trends that best represent these time periods. Each of these eras contain information about important areas of influence including, art, fashion, cosmetics, automotive, architecture, interior design and in later time periods, appliances, toys and electronics.

1 1900s Edwardian Era/La Belle Epoque

Mass-produced fabrics were an offshoot of the industrial revolution that was in full swing by the early 1900s. These emerging materials paved the way for the ready to wear fashion market and softer fabrics and fashions emerged from these new advances in technology. These softened effects worked well with pale and pastel colors that were in sharp contrast to the stiff fabrics and bold colors of the late 19th century. Conversely, this was also an era of very high quality, bespoke fashion that was marked by enormously creative and elegant designs that still influence much of the modern couture market. The Belle Epoque was known as the last age of elegance for the 20th century. Many of these garments for the upper classes were designed with bright colors and a variety of exquisite jewel tones. Summer whites dominated the fashions of all classes of society and were combined with overlays of linen and lace. The Edwardian Era was punctuated with sophisticated shades of celadon, heather, spicy pinks, teal and shades of copper for the masses.

Colored cosmetics were still quite taboo in the Edwardian Era and men were permitted to divorce their wives if they painted their lips so lipstick colors for most women were quite natural and uniform. Colors tended to mimic healthy or bitten lips and these light stains were often derived from natural products such as carmine or carminic acid which is extracted from Cochineal insects. These cactus lice are native to Mexico and the female of species contains large amounts of vibrant red carminic acid which is extracted by squeezing the carapace of the tiny insect. Today carminic acid is also synthetically manufactured and both the synthetic and naturally derived versions are used to color cosmetics, food and many red beverages.

Prior to 1900, automotive colors and technologies were based on the carriage industry and automotive colors varied widely. In 1901, Ransom Olds developed the first modern automotive assembly line and proved its success with the production of the Oldsmobile Curved Dash. Olds built his first steam-driven automobile in 1894 and his first gasoline-powered car in 1896. He also built two different models of

Color Trends and Selection for Product Design.
DOI: http://dx.doi.org/10.1016/B978-0-323-39395-9.00001-3

electric cars around this same time period. In 1903, the Ford Automotive Company was officially incorporated by Henry Ford and cars were hand-painted in red, green and blue.

Although the Art Nouveau era officially started in the 1890s, it was in full swing by the first decade of the 20th century. This "whiplash" style of ornate architecture, like the entrances to the Paris Metro, favored the use of, materials such as steel and glass so colors were in keeping with these materials. Residential archictecture of the early 1900s favored either the lingering Victorian style from the late 19th century or the more austere Colonial style. The Victorian homes favored more lavish decoration with ornate wallpaper schemes and velvet furnishings. Taupe, ivory and deep shades of red, green and accented with golds were typical whereas Colonial homes were more sparse and employed warmer color schemes of deep browns, grayish greens and softened whites. In 1909, Frank Lloyd Wright built one of his last Prairie Style homes, the Robie House, in Chicago. The Prairie Style is closely aligned with the Arts & Crafts Movement but also is distinctively American. Wright designed not only the exterior of the home but also the interior spaces and furnishings thus the color schemes were highly architectural and many neutral, metallics and red brick tones were evident in the furnishings and color schemes throughout. As the art movement of Tonalism was in full swing in this decade, it heavily influenced home décor and design with its emphasis on neutral shades of blue, green and brown awash in a muted or misted effect.

In 1903, Edward Binney and Harold Smith commercialized their first line of Crayola Crayons. The first box of eight crayons contained the traditional ROYGBIV colors along with brown and black. No white crayons were offered. The first box was sold for 5 cents.

In 1908, the first Silver Ghost Rolls-Royce was shipped to India. Off-white with apple green stripes; the Pearl of the East was painted with a lacquer that was rumored to contain crushed beetles to yield a unique luster to the paint.

In 1903, every country that participated in international automotive racing competitions was assigned a unique color. The United States was white and blue, Italy was red, Germany was white, France was blue and Great Britain was green. German cars were also permitted to race in unpainted (silver) to decrease the weight of the vehicle. It is interesting to note that how these colors are still important in the identity of automobiles in these countries today.

Further Reading

http://www.life123.com/beauty/fashion/trends/1900-1910-fashions.shtml
http://www.tudorlinks.com/treasury/articles/view1900.html
https://en.wikipedia.org/wiki/Cochineal
http://www.ideafinder.com/history/inventions/crayon.htm
http://pabook.libraries.psu.edu/palitmap/Crayola.html
https://en.wikipedia.org/wiki/List_of_Crayola_crayon_colors
http://pabook2.libraries.psu.edu/palitmap/Crayola.html
http://pixgood.com/1900-makeup-trends.html
http://ezinearticles.com/?A-Brief-History-of-Makeup&id=516523
http://www.ehow.com/how_12017604_decorate-home-early-1900s.html
https://en.wikipedia.org/wiki/Ford_Motor_Company
https://en.wikipedia.org/wiki/Art_periods
http://auto.howstuffworks.com/1901-1907-oldsmobile-curved-dash1.htm
http://historyofpainters.com/tonalism.htm
http://www.metmuseum.org/toah/hd/geab/hd_geab.htm
http://www.smithsonianchannel.com/shows/the-maharajas-motor-car/0/137811

2 1910s Trends: Arts & Crafts Movement

Although the Arts & Crafts Movement was an international design movement of antiindustrialism that officially started in the 1880s by artists and architects such as the writer William Morris and the designer/architect Charles Ashbee in the British Isles, it grew in strength in breadth throughout Europe and the Americas and was in full swing by the 1910s decade. Much of Ashbee's work was dominated with bright enamels complemented with wrought iron and hand-wrought coppers. The colors that were evident in fashion in the late 1900s decade were also used in the interior of homes where white walls and furnishings were replaced with jewel tones, heavy creams and mossy greens. The Arts & Crafts Movement was an all-encompassed style esthetic and included such items as ceramics, fabrics, metalwork, fabrics and furniture. Many of the designs contained nature-inspired line drawings of plants and animals and flat shades of reds and greens were prominent. William Morris frequently cited that his colors and patterns were inspired by Gothic tapestries but favored natural colorants over the recently developed synthetic dyes that he deemed as garish and more in line with outdated notions of Victorian frivolities and portentous gloom rather than his modern pallet of natural greens, madder reds and dark masculine blue shades.

There were great changes in kitchen design in the first two decades of the 20th century. The great heat waves of the late 19th century underscored the importance of food preservation and kitchen sanitation. Ceramic tiles lined the walls and floors of kitchens and were designed for families, rather than servants. Previously, pantries were used for washing and food storage but during this period, these items and activities became centralized in the kitchens of the bungalow houses that were popular for single, middle class family homes.

As a backlash to the restrictive uses of cosmetics in the previous Edwardian Era, Suffragettes fought for their right to wear lipstick and take charge of their appearance and chose red lipstick as a symbol of their independence. Lip and rouge colors were vibrant and in sharp contrast to very pale almost theatrically white foundation. 1915 is the

Color Trends and Selection for Product Design.
DOI: http://dx.doi.org/10.1016/B978-0-323-39395-9.00002-5

first date of commercial lipstick in the United States when a French cosmetician, Anthony Guash and Maurice Levy, a NY investor, formed the French Cosmetic Manufacturing Company in Connecticut. Prior to this, women would mix dyes to create their own formulations obtained at local pharmacies.

In 1917, Cutex offered their first tinted nail enamel in a rose shade of pink. This was the first evidence of colored nail enamels as a fashion statement in modern society but nail decoration was certainly not a new concept. Ancient Babylonians and Egyptians decorated their nails as an indicator of social status. Babylonian, middle class males wore green nails and upper class males were decorated with black nails. The pigmentation was created from grinding kohl, antimony sulfide and a grayish black salt that oxidizes to a jet black upon exposure. This was not the most health-friendly cosmetic since its toxicity is similar to arsenic. Queen Nefertiti also decorated her nails with gold and a Henna extract to create a red-iron oxide color. Lower class Egyptian women were only permitted to use pastel colors on their nails.

This was also the era of Brass Automotive vehicles whose many fittings were comprised of brass but colors were mostly relegated to shades of black or white. In 1913, the concept of "Fordism" or mass-production of automobiles was officially born and was also the advent of the Japan Black lacquer paint that dried quickly and was much easier to apply than earlier varnishes and lacquers that were hand-brushed on wood and metal surfaces. Ford selected black as the color of choice for the sake of efficiency and durability. However, there were many automotive companies around the world that offered numerous styles and colors but Ford's mass-produced black varnished vehicles were most common. The George V coronation in Delhi in 1911 represented the zenith of imperial rule in India. Rolls-Royce exported eight silver ghosts for the coronation ceremony and the eyes of India's most wealthiest citizens were upon the Rolls-Royce vehicles vying for future purchases.

In 1918, there was an article written in Earnshaw's Infant's Department, a trade publication which stated that pink was a perfect color for boys because it is a stronger and more determined color and blue was better for girls because it was viewed as a more delicate and flattering color for blonde haired, blue eyed babies. Other sources felt that pink was better for brunette babies with brown eyes, regardless of gender. In 1927, Time Magazine published a chart confirming this pink for boys and blue for girls idea and the trend continued until the 1940s when the industry looked to consumers to see how they implemented these ideas and realized that most consumers preferred to dress their boys in blue and their girls in pink and then the manufacturers changed their color schemes to compensate for consumer trends.

French artist Henri Matisse was an important artist and color influencer of this era and also in the 1990s after his retrospective in the MOMA, New York. In both instances, his use of bright colors, strong reds, balanced with bright and passive blues, turquoise and chartreuse combined with natural yellows and oranges, influenced home and fashion design in the years following his exhibitions. His use of layering and lining strong, arbitrary colors shocked the established art world that called his work beastly or *bête sauvage* led to the formation of the Fauvist art movement.

Further Reading

https://en.wikipedia.org/wiki/Charles_Robert_Ashbee
http://www.ehow.com/list_7154070_decorating-tips-1910-home.html
https://en.wikipedia.org/wiki/Arts_and_Crafts_movement
http://glamourdaze.com/history-of-makeup/1900-1919
http://www.smithsonianmag.com/arts-culture/when-did-girls-start-wearing-pink-1370097/?no-ist=
https://blog.kissmetrics.com/gender-and-color/
http://web.archive.org/web/20140415055035/http://www.npr.org/blogs/thetwo-way/2012/12/17/167452439/girls-boys-and-toys-rethinking-stereotypes-in-what-kids-play-with
https://en.wikipedia.org/wiki/Japan_black
https://en.wikipedia.org/wiki/Brass_Era_car

3 1920s Trends: The Roaring Twenties and the Art Deco Era

After the end of World War I, an economic boom was evident in many parts of the world. The development of new synthetic colorants and materials such as Rayon (synthetic silk) was explosive in the 1900–30 time period and most of the colors that were developed prior to this period were rarefied versions of naturally derived colorants. Clothing became more comfortable and sportswear was born. The era of the "Flapper" began with more relaxed morality codes and personal freedom was evident in clothing styles and colors. Zelda Fitzgerald, wife of F. Scott Fitzgerald, was the embodiment of this Flapper style and a muse for the Jazz Age. Zelda was frequently seen in short flared skirts in shades of orange, red, green and gray. In 1926, American Vogue featured Coco Chanel's "Little Black Dress" and called it the "Little Boy Dress." The first version was constructed of silk and had long sleeves. It was calf-length and quite straight. The dress was also very fashionable and comfortable. It was stripped away of excesses and more practical than other designs of this time period. This dress has been fashioned for day wear and evening wear in different fabrics, always remaining black but still an enduring wardrobe staple. In contrast, the Liberty of London pallet of the 1920s shifted from dusky and somber colors to a much brighter pallet of bright blues, yellows and reds similar to Wassily Kandinski's colors of the same time period. Liberty of London was a textile manufacturer that started in London in 1875 and was inspired by colors of Asian fashion and art. This Asian Exotica was introduced to the British middle class and their peacock feather motif was wildly popular in the 1920s. There was also a subtle shift in coloration from the Arts & Crafts influence of the early 1900s with their dark hues, shifting to Liberty's grayed colors as a sharp contrast to coral, pink and bright greens. Wassily Kandinski is sometimes referred to as the Godfather of Modern Art who celebrated the emotional content of color in his paintings.

In 1927, Time Magazine published an article indicating gender-appropriate colors for boys and girls. This information was based on

Color Trends and Selection for Product Design.
DOI: http://dx.doi.org/10.1016/B978-0-323-39395-9.00003-7

marketing strategies by major US retailers such as Marshall Fields in Chicago and Filene's in Boston who suggested that shades of pink were more appropriate for boys and blues were better for girl's clothing. This was reversed in the 1940s when consumers forced today's preferences of pink for girls and blue for boys through their buying patterns.

After the war, an abundance of scrap metal allowed for the development of metal lipstick tubes, replacing the fragile cardboard or paper tubes that were dominant in the United States and Europe. Lip and nail colors of the 1920s were very dark. A very popular color was an aubergine or dark brown eggplant color, even black was used as a contrasting color for the black and white film industry to accentuate the shape of the lips and differentiate from lighter skin tones. However, off the screen and stage, many women opted for lighter, more natural shades and it was during this era that the still-popular "Tangee" lipstick was developed. This was developed as a novelty formulation containing Eosin that appeared orange prior to application and then shifted to a moderate shade of pink or even more of a seductive shade of red depending on the pH of the wearer. Tangee was very critical of the painted look of some of its competitors and appealed to a more wholesome clientele. Initially, they offered only two shades: "Natural" for everyday wear and "Theatrical Red" for those in theater industry. In 1926, Helena Rubenstein developed the "Cupid's Bow" lipstick applicator in shades of Red Raspberry for medium complexions and Red Geranium for lighter skin tones or blondes. The lipstick tip was molded into a V-shape that allowed for perfect application to the natural shape of the vermilion border. Additionally, in 1927, Paul Baudecroux introduced a Black Magic lipstick in France that appeared black in the tube and transitioned to pink or red after application. This change is due to the use of Eosin in the formulation.

In 1917, there were 127 automotive manufacturers in the United States. This is known as the Vintage Era of automobiles. The competition in the automotive market was brisk and manufacturers realized that they needed to compete for the attention of women drivers. The marketing strategy designed more comfortable, plush upholstery, interior heaters and colors other than Henry Ford's black.

Despite the decline of luxury cars because of World War I, many Rolls-Royces were imported into India in the 1920s by the Maharajahs and were known as the mechanical elephant. Chassis and engines were delivered to the coach builders where the body was custom-designed for their owners, by such companies as Windovers in London in luxurious shades of deep blue, hunter green and burgundy. In 1921, Rolls-Royce decided to commission several prominent coach builders to design all of the bodies for their automobiles and the fully manufactured vehicles were sent to the new showrooms that had opened throughout India. Rolls-Royce also sent drivers and mechanics to service the vehicles for their owners and cemented their loyalty with India's elite. In 1925, Rolls-Royce introduced the Phantom 1 that was used as a tiger-hunting vehicle that was clad in stainless steel and brass for durability.

In 1923, General Motors introduced a vehicle painted with a new nitrocellulose-based paint from Dupont called Duco. This fast-drying, easy to color paint revolutionized the automotive paint industry by decreasing painting time from 2 weeks to 2 days. At this time, Ford also expanded their color pallet to a wider range of colors on the Model T and the subsequent model A. Most of these were darker shades of green, blue and red but also continued with the classic black. Many two-toned vehicles entered the market and a new era of colorful automobiles graced the emerging highways. The Grant touring cars were painted in exquisite shades of blue and turquoise and were so esthetically appealing that they were exported to England as Whiting-Grant vehicles. Luxury vehicles were oftened distinguished from commodity cars, in part, by the use of tan and cream paint schemes.

Single-family homes were smaller in the 1920s than in earlier decades. Two popular styles were the bungalow and the four-square. Most of these were wood-frame structures and were painted with a variety of subdued, greens, grays, browns and brick reds with a contrasting white trim.

The Art Deco movement originated in the 1920s and was popular for many of the following decades. It was inspired by Cubist art that favored repetitive lines and clean geometric patterns that developed in the previous decade by artists such as Pablo Picasso and Georges Braque. The Expressionist Architecture movement was also an important design movement of the 1920s and emphasized architecture as an art form rather than a primarily structural edifice. The Expressionist style was more elemental and geometric whereas the Art Deco style was more floral and organic in volume and tone.

In 1927, General Electric revolutionized kitchen sanitation with the introduction of the Monitor Top Refrigerator which was an hermetically sealed refrigeration unit atop of a moderately sized rectangular fixture. The name "Monitor Top" was attributed to the protruding compressor that looked similar to the turret of the warship; the USS Monitor. Most of these refrigerators were painted in pure white or off-white shades. Other shades of reds, yellows and blues were introduced in the 1930s.

Further Reading

https://www.pinterest.com/perlada/
https://en.wikipedia.org/wiki/1920s
http://fashionthroughtimeuonkt.blogspot.com/2012/04/1920s-youth-culture.html
http://www.1920-30.com/automobiles/
http://www.ehow.com/info_8448414_exterior-colors-1920s.html
https://en.wikipedia.org/wiki/1920s_in_Western_fashion
https://en.wikipedia.org/wiki/Little_black_dress
http://www.listal.com/list/fashion-icons-every-decade
https://en.wikipedia.org/wiki/Lipstick
http://www.smithsonianmag.com/ist/?next=/arts-culture/when-did-girls-start-wearing-pink-1370097/?page=1
http://www.ehow.com/list_7451497_color-schemes-1920s.html
http://www.searsarchives.com/brands/kenmore.htm
http://www.antiqueappliances.com/monitor_top_refrigerators.htm
http://miniatures.about.com/od/referencesandbookreviews/tp/References-For-Modelling-Miniature-And-Dollhouse-Kitchens.htm?utm_term=hoosier%20cabinet&utm_content=p1-main-2-title&utm_medium=sem&utm_source=msn&utm_campaign=adid-32f45af9-2141-45ca-8665-6c44b56388c9-0-ab_msb_ocode-7931&ad=semD&an=msn_s&am=broad&q=hoosier%20cabinet&dqi=hoosier&o=7931&l=sem&qsrc=999&askid=32f45af9-2141-45ca-8665-6c44b56388c9-0-ab_msb
http://www.cosmeticsandskin.com/ded/lipsticks.php
https://en.wikipedia.org/wiki/Grant

4 1930s Trends: A Golden Age of Growth-Bauhaus, Art Deco and De Stijl

Although the German Bauhaus, movement of modernist architecture started in 1919 and ended in 1933, its greatest era of influence was clearly in the 1930s and is an example of how intellectual schools of thought predate actual industrial and manufacturing trends. Bauhaus was quite unique in its attempt to unite art, architecture and design, including typography, into one esthetic movement. The Bauhaus color pallet was quite muted and earthy that ranged from a pallet of blues to browns and oranges to dark greens and pale yellows. All of these colors were quite subdued in nature and reflect how the psychology of nature reflects the spirit of space and composition in architecture. Color proves to be key for this composition. Rich and full colors not only rhythmatize and structure the arrangement, but also represent pictorial values in their own right, which are not integrated in a homogeneous whole. Conversely, Piet Mondrian's artistic movement, De Stijl (The Style) or Neoplasticism, was also born in the earlier 1900s but emerged in the 1930s a defining study in color contrasts employing the primaries, red blue and yellow versus three values of black white and gray. It is generally thought that De Stijl artists were highly inspired by medieval stained glass patterns but the strict primary colors were also in response to the violence of the World War I. In 1937, Surreal artist Salvador Dali and fashion designer Elsa Schiaparelli presented surrealistic fashions including the Shoe Hat which was a stiletto-shaped black felt hat with a shocking pink heel. Marlene Dietrich and Wallis Simspon were two style icons of the 1930s for very different reason. Marlene Dietrich defined her own style esthetic, wearing exquisite white gowns and black evening wear as well as men's clothing in shades of grays and blues. Wallis Simpson was the wife that George VIII chose over his British crown and her style was classic and sophisticated. On her wedding day, she wore silk gown that was styled in a sapphire color that was dubbed, "Wallis Blue" because the color was matched to her eyes. She also wore light blue sueded pumps to complete the ensemble.

Color Trends and Selection for Product Design.
DOI: http://dx.doi.org/10.1016/B978-0-323-39395-9.00004-9

By the 1930s, over 90% of urban households were wired for electricity in the United States. This caused a surge of new electrical appliances, including washers and dryers, to appear in American kitchens mostly in shades of white but that also shifted into more highly chromatized shades after the World War II. Similarly, kitchen colors were still mostly pure white but for the wealthy there were also emerging colors in vibrant shades of bright yellow, spicy red, royal blue, pink and botanical greens complemented with patterned or checkerboard flooring and wall coverings. Unlike the 1920s kitchens, cabinetry was mostly built-in steel with white porcelain sinks rather than the free-standing Hoosier cabinets that functioned as work and storage space in the kitchen.

Despite poor economic conditions, lipstick sales soared in the United States. This is a trend that often accompanies difficult economic times where microluxuries flourish as a coping method to counteract larger difficulties. Another reason lipstick use flourished in the 1930s is because it feminized the militaristic styles of clothing which were prominent in this era. In the 1930s, Max Factor invented the gloss lipstick and lipstick colors start lighten to a more red-shade purple. In 1935, Cutex, the nail polish company, introduced the self-serving idea that nail enamels should match lipstick colors and issued four new shades of lipsticks to perfectly align with their nail enamel colors: Natural, Coral, Ruby and Cardinal.

Automotive colors became brighter shades of red blue and green but the Great Depression hit the automotive industry very hard and many luxury brands and manufacturers disappeared along with the cream and tan shades of automotive paint.

Ferdinand Porsche invented the VW Beetle in 1934, creating the ultimate people's car. Adolf Hitler asked him to design the German version of Ford's ubiquitous Model T. Hitler then tapped Porsche to make tanks for World War II and the production of the VW Beetle's ceased until 1949 in Germany when it was imported into the United States.

Although polyamide was invented by Wallace Carothers at Dow in 1927, it took Dow 10 years to formally introduce nylon to the commercial market. In 1939, Dow began researching and replacing silk parachutes with the new nylon material. After the war was over, these nylon parachutes were recycled into the clothing market for dresses and stockings.

Further Reading

http://www.lipstickhistory.com/lipstick-history/timeline-of-lipstick/
http://www.twoop.com/archives/2005/10/lipstick.html
http://www.metmuseum.org/research/libraries-and-study-centers/in-circulation/2014/architecture-vitrine (1930's Bauhaus).
https://artblart.com/tag/color-theories-of-the-bauhaus/
https://en.wikipedia.org/wiki/Bauhaus
http://www.ehow.com/info_7963187_differences-art-nouveau-edwardian-victorian.html
https://en.wikipedia.org/wiki/De_Stijl
http://glamourdaze.com/2013/02/a-brief-history-of-eye-makeup-by-gabriela-hernandez.html
http://www.dupont.com/corporate-functions/our-company/dupont-history.html
https://en.wikipedia.org/wiki/Wallace_Carothers
http://www.metmuseum.org/collection/the-collection-online/search/81119
http://www.ehow.com/info_8361375_kitchen-colors-1930s-1940s-1950s.html
https://www.youtube.com/watch?v=AINcSU-kJ28 (Pour un soir – 1931 – Jean Godard – Les Arcades du Lido – Paris).
http://www.protectall.com/artpaints.aspx
http://www.visual-arts-cork.com/famous-artists/bridget-riley.htm

5 1940s Trends: Global Military Struggles

Colors and color trends in the 1940s were very much influenced by the global world war. After the fall of France in 1940, fashion was largely driven by Hollywood in the United States where the birth of Film Noir captivated the fashion and color world. Minor influences were also evident from London during this time period.

In the first half of the 1940s, many parts of the world were focused on providing materials and technologies related to the war effort and less attention was paid to the artistic and esthetic sides of society. Lusterless Olive Drab Enamel was a US Government standard for all military vehicles (including tanks) that started in the 1940s and continued until the development of camouflage patterns in the 1970s. This also resonated well with the earlier Art Deco colors of the 1920s for interior furnishing and textile colors in the 1940s that were popular in shades of softened grays combined with fiery reds, bright yellows and pistachio greens.

Fashion trends tended to be militaristic in both design and color. There was an excess of tailoring and belting complemented by large shoulder pads that suggested strength and stress. Since there was a limited amount of materials available, the US government directly imposed restrictions on the fashion industry; banning pleating and excess folds and ruffles. Restrictions were also enforced on lengths of skirts, sleeves and dresses and the width of belts. These restrictions heavily influenced the fashion industry both in terms of style and color. Men's suits were cut quite slim and women's hem lines were so high that liquid eyeliner was used to create a pseudo seam up the back of a bare leg that a stocking might have been covered if the parachute industry had not absconded with all of the silk. The colors were quite plain and since green and brown dyes were reserved for the military, natural whites and beiges dominated the fashions and were combined with various shades of grays and dark burgundies and red wine colors. Floral prints were important during the start of the 1940s, with the mid-to-late 1940s were dominated by geometric or modern prints. In 1940, Elsa Schiaparelli introduced her spring collection that included camouflage printed taffetas.

Color Trends and Selection for Product Design.
DOI: http://dx.doi.org/10.1016/B978-0-323-39395-9.00005-0

The second half of the 1940s were marked by more luxury in fashion styling where dresses became longer and were more elaborated with pleats and ruffles. Men's suits were more generous in fabric, larger shoulder pads, wider lapels, trousers and double-breasted suits emerged.

In 1946, Louis Reard introduced the first Bikini bathing suit. Reard could not find a conventional model to wear the costume so he hired a nude model for the photographs. The two-piece suit was so small that the model was photographed holding a small box for matches in which the suit could fit inside. Reard named the suit after Bikini Island where nuclear bomb testing had occurred since Reard felt that all women who wore these bathing suits would be regarded as bombshells. The French culture embraced this new fashion statement but the Bikini was banned in many states in the United States as well as many countries in Europe including the heavily catholic influenced Italy. Another postwar fashion trend was the Hawaiian print shirt that was cemented in popularity when US president Harry Truman was spotted wearing this highly flamboyant style of men's shirt. The Hawaiian Islands came into focus after the Japanese attack on Pearl Harbor and some felt that these large floral prints were a sign warning coloration and a show of strength by American citizens. These prints were popular in blood red, ocean blues, oyster whites, punctuated with tropical oranges and pineapple yellows.

In 1947, Christian Dior felt that the world was ready to depart from the sharp-shouldered suits and the knee-length skirts that dominated the earlier part of the decade introduced the "New Look" to the Paris fashion world. This "Flower Woman" look exhibited waspy waists, soft shoulders and luxurious flowing skirts and exemplified an accomplished feminine woman who was leaving the factory to return to care for her home and family. However the color pallet remained largely untouched, still utilizing many gray and burgundy tones. In 1948, French designer Jacques Fath introduced a majestic black swing coat with a stunning pink lining that was a perfect complement to the full skirts and the swells of pregnancies that were prominent in the United States and Europe in the postwar era. In 1949, Fath designed a red and green bathing ensemble that included a large hat and modesty wrap that used unique dyes, that made his fashions vibrant and unrivaled by his competitors. Ingrid Bergman and Hedy Lamar were fashion icons of the 1940s. Ingrid Bergman was the embodiment of Dior's "New Look" whereas Hedy Lamar was stunning and glamorous in many of her movies wearing shades of Orchid, Bright Blue and Crimson.

In 1945, the first Fashion Week was held in New York to detract from French design houses during World War II since many fashion professionals could not travel during this time. Until this time, the French fashion world was responsible for inspiring the rest of the world but this "Press Week" in New York dramatically shifted the media attention from Paris to New York and other parts of the world.

After the close of World War II, the art world flourished and the overlapping movements of Modern and Contemporary Art emerged. Color Field Painting emerged as part of the Abstract Expressionism art movement in New York City in the late 1940s and grew in popularity in other parts of the world in the 1950s. This type of painting is characterized by large areas of flat, unbroken color without a central point of focus but rather the canvas is viewed as an entire field of color. This cerebral type of painting is less focused on the physical act of painting or brushstrokes employed but rather emphasizes the emotional expression that is created by layering or overlapping the colors and is more closely related to fashion as a result and was certainly an influence in fashion color trends in the 1950s and 1960s. Mark Rothko's large rectangular paintings are a prime example of Color Field Painting, although Rothko feels that his color selection is secondary to the emotional content of his paintings, his color choices are layered and sharply contrasting patterns of bold saturated colors of orange, pink, bright blue and green with moody grays, rust and dark blues.

The domestic color pallet became more feminine and vibrant and the use of dyes like Dylon and RIT was used to brighten up fading homes and older fashions of the masses. The Color Helm by Ostwald was a color wheel that was used to guide home makers and fashionistas to select the correct combinations of colors that were appropriate for the late 1940s.

As William Morris rebelled against the overuse of bright synthetic colorants in the earlier 1900s, Frank Lloyd Wright also took issue with the mass-produced machine age that was invading home interiors and architecture. Wright embraced natural colors, textures and materials and created interior color pallets that harmonized with the natural exterior structures of his buildings where he favored natural wood and brick colors and complimented them natural tints of bright greens, lemon and slate blues.

At the same time, the Art Deco Architectural movement was growing. This trend started in the 1930s and became wildly popular in Miami, Florida as a marker for the end of the depression and is sometimes referred to as the Modern Depression architecture that featured a pallet of tropical colors in ocean greens, tropical sunny yellows and coral pink tones.

The overuse of petroleum products for the war effort created a scarcity of lipsticks for the consumer. Metal tubes were replaced by mostly by paper. Lip and nail colors of the 1940s were bright, matte colors of stinging corals (Granada) and blood-red, almost angry colors directly related to the global military conflict and an offset to the drab colors of the fashions. In 1941, The Du Barry cosmetics company launched their Emblem Red color, stating that their blood-red shade was crisp, daring and military for the spring collection.

Automotive colors became darker as well as manufacturers tended to make more conservative, "Black Out" vehicles with little or no chrome and almost exclusively black paint. Ford offered 13 different colors in 1941 and these choices swelled into the dozens by the 1950s. Citroen's Traction Avant drastically reduced their color choices to black with tan or red wheels.

In 1949, Crayola issued their first white crayon to use for blending with other primaries to create pastel shades.

Further Reading

http://design.designmuseum.org/design/christian-dior
https://en.wikipedia.org/wiki/1940s
https://en.wikipedia.org/wiki/Elsa_Schiaparelli
http://www.ehow.com/info_7992506_school-uniforms-1940s.html
http://olive-drab.com/od_mvg_odpaint.php
http://www.ehow.com/info_8792398_paint-codes-1940-ford-cars.html
http://content.time.com/time/specials/packages/article/0,28804,2049243_2048651_2049010,00.html
https://en.wikipedia.org/wiki/Color_Field
http://news.discovery.com/history/us-history/bikini-history-110701.htm

6 1950s Trends: Automotive

The rising postwar prosperity fueled the true love-affair with everything automotive that started and peaked in the 1950s. Car-starved consumers were looking for futuristic, gadget-filled vehicles, building on the new technologies that were developed for wartime. The 1950s automotive colors and designs were unusually flamboyant; ranging from turquoise to chartreuse, magenta to robins-egg. These colors were frequently combined with excessive chrome accents as well as dramatic fins and sumptuous curves. Ford offered 38 different paint colors by 1950. In 1955, Popular Mechanics surveyed over 1000 Ford owners and the biggest appeal of their vehicles was their color and appearance. Some owners felt that the brighter colors were safer and provided greater visibility on the road. Others felt that the colors made the car and driver feel younger. Bright colors were also thought to hide the dirt better and other owners felt that the brighter colors were more modern. However, not everyone liked the bright colors; some customers felt that these colors were too loud and others were worried about the paint colors fading over time. Some customers felt that there was too much chrome and complained that it created too much glare for the driver.

Although the stoving enamel spray paints in the 1930s were glossier, a new reflow acrylic was developed in the 1950s. This paint was also sprayed onto the vehicle but then cured in an oven at 350°F to drive off the solvents and yielding a very smooth paint surface. However, these acrylics were not as glossy as their alkyd predecessors but the painting process was much quicker and more efficient.

Although Packard started painting their vehicles with brighter colors in the late 1940s, the colors became much more vibrant by the mid-1950s. An entire series of Caribbean blues and greens, reds and yellows were developed and were complimented by a pallet of fuchsia, teal and turquoise balanced by golds, silvers and bronzes.

Two-car families were a new phenomenon that emerged in the 1950s. This allowed automotive manufacturers to create more whimsical, less practical designs as well as marketing directly to many women who had careers during the war and had a much greater influence in purchasing products for the family.

Color Trends and Selection for Product Design.
DOI: http://dx.doi.org/10.1016/B978-0-323-39395-9.00006-2

The 1950s was the first era of gender-specific cars. In 1954, Chrysler created the Le Comte and the Le Comtesse vehicles. Although they were based on the same Chrysler Newport vehicle, the Le Comte was painted in masculine colors such as bronze and black and the La Comtesse was decorated with a pink and dove gray color scheme. In 1955, Dodge introduced an exclusively feminine car known as the La Femme. The color scheme was "Sapphire White" and "Heather Rose" and the La Femme model was created in gold script by the front fender. The chromed headlight surrounds were heavily lidded, and tapered to a fine point on the sides of the vehicle reminiscent of a Cleopatra eye mascara. The interior was designed with pink rosebuds on a silver-pink background with pale pink vinyl trim. The dashboard was a black over pink color scheme and the very 1950s black and white steering wheel was mounted on a pink steering column. There were two storage boxes in back of the front seats. The box in back of the driver's seat contained a raincoat, rain hat and umbrella that matched the jacquard rosebud interior. Chrysler marketed the car by stating that it was made "By Special Appointment to Her Majesty... the American Woman." The car also came with an elegant pink leather purse that matched the interior of the car. The carry-all shoulder bag was stowed in a custom area in the box in the back of the passenger seat, that had a gold metal plate that faced outwards and with the owner's name engraved on it. The purse contained many items essential to a woman's comfort: a cigarette case and lighter, a tube of lipstick, face powder and comb that were coordinated in gold metal with pink leather that matched the interior of the vehicle. The entire ensemble was designed and manufactured by Evans, a prominent designer in the 1950s. In 1956, the color scheme of the La Femme graduated into a more sophisticated scheme of Misty Orchid and Regal Orchid colors. The seating was white and decorated with a random lavender pattern trimmed in dark gold. The headliner was also white with specks of gold while the carpeting was designed with several shades of lavender and purple. Although the marketing department called the La Femme a, "stunning success," less than 2500 of these vehicles were ever created and its failure was attributed, in part, to a lack of promotional advertisements when compared to their more masculine counterparts.

In 1957, the space age officially started when the Soviet Union launched the Sputnik that orbited around the earth but American Automotive manufacturers created cars with large tail fins that were designed to look like space-age rockets that were ready to go into orbit that could circle the earth.

In 1958, Edsel introduced the Bermuda Station Wagon that combined synthetic wood paneling along with exotic colors such as Ember Red,

Chalk Pink, Ice Green and Jonquil Yellow as well as the ubiquitous Turquoise.

Prior to the 1950s, most kitchen appliances were white or off-white but in 1950s colors such as pastels in the shades of yellow, green, turquoise and blue, natural browns, pinks and bright yellow helped consumers coordinate their home furnishing colors and closely resembled the emerging automotive colors. Similarly, after the World War II, Sears introduced the "Harmony Home" line of Appliances in Sunshine Yellow, Malibu Coral and Aquamarine. Also, office colors became brighter. Black and gray office equipment shifted in softer shades of ecru and pink that was friendlier to the emerging female workforce that descended upon the offices and factories of the Western world.

The mid-century modern houses of the 1950s to the 1960s were influenced by the explosive growth of suburban communities, populated by men returning from the World War II who wanted to start their families and careers. These homes featured open floor plans with ample windows and were quite modest in size but immodest in color: jade, blue gold and coral exterior paint colors accented with off-white trim were quite common.

Mimo Architecture, also known as Miami Modern Architecture, fluorished in Southern Florida in the 1950s. The color schemes were similar to those in the modest mid-century modern homes but with their dramatic, space-age architecture, they were far from modest in the appearance along the gold coast Atlantic. The most iconic structure is the Fontainebleu Hotel in South Beach that built in 1952 from the Harvey Firestone Estate which had fallen into a state of disrepair over the last 20 years and was recently revived in 2008.

The 1950s were also known as the era of the "Lipstick Wars" where Hazel Bishop and Revlon mounted highly ambitious marketing campaigns. The Bishop products celebrated the practical woman with sensible colors whereas the Revlon products targeted the more romantic consumer using celebrities such as Marilyn Monroe with her platinum hair and blood red lipstick and Elizabeth Taylor's Cleopatra character with her dramatic blue eye shadow with a super light peachy coral gloss. Generally, nail and lip colors were sweeter and more wholesome, much pinker with a slightly blue shade. The prosperous economy also allowed for more luxurious packaging for cosmetics and prices varied widely. In the United States, Revlon lipsticks ranged from a little over $1 up to $35 for cases fabricated in sterling silver. The upper classes could take their Revlon lipsticks to a jeweler and have custom cases designed with precious metals and gems. After the destruction of

Europe from World War II, consumers looked to Hollywood for fashion ideas rather than the runways of Milan and Paris.

The 1950s were also a pivotal time for youth style. Prior to the 1950s, 18-year-old women of Europe and the United States dressed in the same fashions as their mothers for whom the styles were intended. The term teenager was also coined in the 1950s and separated this consumer youth group from boys and girls to adult men and women. American teenage boys favored the Preppy and Greaser style esthetics whereas the British teenagers developed their Teddy Boy look. The Greaser style was embodied by the James Deane and Elvis influences. The fashions were simple and consisted of blue jeans and black leather motorcycle jackets. As the British economy improved in the 1950s, young men desired a more tailored, Neo-Edwardian custom suit style that separated them from the shabby styles of their fathers. This was a fashion style that was developed by the working class youth on the streets of London rather than the established tailors on Saville Road. The whimsical Teddy Boy was a derivation of the Edwardian Style esthetic from earlier in the 20th century and was characterized by velvet-trimmed long jackets and were typically fashioned in tone-on-tone shades of claret, dark blue and black. The Teddy Boys also wore crepe soled shoes perfect for dancing to the beats of the 1950s music. The Preppy style from the United States for teen boys was characterized by white socks, suede shoes, clean and crisp light-colored button-down shirts, pleated trousers and navy blazers. The teenage girls counterparts of these 1950s fashion trends were just as interesting. The Teddy Girls wore low-cut sweaters, usually in black, along with circle skirts or Capri pants in salmon, turquoise, navy blue and cream. The Teddy Girls usually pulled their hair back in pony tails. The Greaser Girls were more practical in their fashion choices. Their hair was usually pulled back, in a bandana or retained by a headband. Blue denim capris or shorts were very popular along with button-down shirts tied in a knot at the belly button, red and white checked or paisley prints were popular. Preppy Girls were clean and wholesome in appearance and employed color schemes of clean pastels to complement their male counterparts. These were colors of optimism, prosperity and growth. Many color names had flower references that pointed to fertility and promise of beauty including many peach and blush colors. Poodle or full dirndl skirts with large appliqués of poodles, cars or flowers in pink, white or black on their clothing were very popular. Neat pleated polyester skirts with sunray pleating were also prominent in pinks, blue, white, teal and coral. The skirts were lined with nylon or synthetic mesh underlayments. Teen Preppy Girls also wore brightly colored scarves that were tied at the side of the neck, under scoop neck blouses, button-down cardigans or three quarter length sleeve white button down blouses.

Marilyn Monroe and Audrey Hepburn were two of the most iconic fashion stars of the 1950s. Monroe was characterized as sultry and effortless. In 1953, Monroe starred in "Gentlemen Prefer Blondes" and wore a pink gown while singing "Diamonds Are a Girl's Best Friend." Another fashion classic was her white short blowing dress in "The Seven Year Itch." Audrey Hepburn was almost the complete opposite of Marilyn Monroe, she avoided trendy clothing and was well suited by classic fashions and colors frequently spotted in black turtle necks and simple white dresses. 1950s glamor was solidly defined when Chanel re-entered the fashion stage in 1955 with her classic Chanel Suit, produced with brightly colored Scottish Tweeds.

Binney and Smith had two significant developments in the 1950s. In the spring of 1950, they introduced Silly Putty, a silicone product marketed as a children's toy. The product was released on the market as an Easter promotion and was appropriately packaged in plastic shaped, spring-colored eggs. In 1959, Crayola also introduced their "Cadillac of Crayons" a 64 crayon color assortment box with a built-in sharpener.

Two important art movements of the 1950s were the Pop Art Movement and Post-Modernism. The Pop Art Movement developed in Britain in the mid-1950s and in the late 1950s in the United States. Pop Art is characterized by an appreciation for popular cultures that were prominent in the 1950s such as comic books, popular media and advertisements for goods and services. Two of the most celebrated British Pop Artists were Peter Blake and Richard Hamilton. Two of the most celebrated American Pop Artists were Andy Warhol and Roy Lichtenstein. The style of the movement employed very bright, flat colors, pixelated or Ben-Day dots that were used in Roy Lichtenstein's work. Stencils were frequently used to create multiple images of everyday objects and celebrities. These techniques were prominent in Andy Warhol's work when he produced the multiples of Campbell's Soup Cans and Marilyn Monroe.

The artist Jackson Pollock also rose to prominence in the 1950s. Although Pollock started creating his Action Art in the 1940s, his style of actively dripping and spattering large canvases that were frequently laid on the floor rather than on an easel became highly celebrated with the help of Peggy Guggenheim in the 1950s. The earlier Action Art of the 1940s were brighter and more optimistic in color and became progressively darker and more dense in the 1950s and is frequently attributed to Pollock's sustained addiction to alcohol but could be simply an evolution of his expression that was independent of his condition. Jackson Pollock's work is considered by some critics as early fractal art due to its repetitive and random nature.

The Postmodern Art movement developed after the end of World War II as a reaction to the failing of artistic society becoming entirely too mainstream, the atrocities of the Holocaust and was later fueled by such global events as Vietnam and the Cuban Missile Crisis. These types of events instilled a sense of disillusionment of the meaning of life and art. It is characterized by having a tenuous relationship to traditional art forms that had difficulty establishing a definition of real art. The mistrust of artistic ideologies is an essential component to its identity.

Further Reading

http://www.colorcombos.com/kitchen-appliance-color-choices-article.html
http://boblemke.blogspot.com/2014/11/world-on-wheels-custom-1955-la-femme.html
https://en.wikipedia.org/wiki/Dodge_La_Femme
http://www.allpar.com/model/lafemme.html
https://en.wikipedia.org/wiki/Miami_Modern_architecture
http://www.packardclub.org/
http://edwardianteddyboy.com/
https://books.google.com/books?id=xdwDAAAAMBAJ&pg=PA133&dq=1954+Popular+Mechanics+January&hl=en#v=onepage&q=1954%20Popular%20Mechanics%20January&f=false 1950's
http://www.crayola.com/faq/another-topic/why-does-the-color-quotfleshquot-not-appear-in-the-1958-limited-edition-box-of-64/
https://en.wikipedia.org/wiki/Ben-Day_dots
http://arthistory.about.com/cs/namesww/p/warhol.htm
https://en.wikipedia.org/wiki/Pop_art
http://arthistory.about.com/od/modernarthistory/a/Pop-Art-Art-History-101-Basics.htm
https://en.wikipedia.org/wiki/Jackson_Pollock

7 1960s Trends: The Mods and the Psychedelic Era

The 1960s were a decade of change and experimentation and color trends closely mirrored this phenomenon where the early 1960s started with dark and muted tones but increased in clarity and hue that continued well into the 1970s. Part of the reason colors became brighter in fashion and design related to the light fastness and stability of these highly chromatized shades. This enhanced feasibility inspired designers to become more imaginative and creative with their designs. The 1960s were a prosperous decade of many new fashion trends. Some of these were influenced by social movements and others were influenced by celebrities such as Jackie Kennedy with her pillbox hats, and boxy, Chanel style jackets with large buttons and pencil skirts. In the early 1960s, the London Modernists, or Mods, were overriding fashion designers and dictating the current styles and fashion trends. Mary Quant was a pivotal designer for this movement and in 1963 she received the Dress of the Year award by the Bath Fashion Museum for her mini dress that was fashioned in a cream wool jersey with blue trim. She saw fashion as more of a whimsical game than an industry and created fashions that encouraged young people to dress to please themselves.

Young men were following the hair fashion styles worn by rock bands such as the Who and the Beatles that favored close-fitting jackets and pants as well as Edwardian collarless silver-gray suits with moptop-style haircuts. These styles were, in part, an evolution of the Teddy Boy style that was popular in the 1950s. Women wore very short skirts, the minis and the micros as well as brightly colored go-go boots and sleeveless dresses and blouses. These fashions emulated styles more associated with children than adults.

In 1966, Andy Warhol created his paper "Souper Dress" that was marketed and sold by the Campbell Soup Company for two soup can labels and 1 USD. The idea of paper dresses and disposable clothing was a fad that developed in the 1960s and was short-lived due to their ill-fitting, uncomfortable and flammable characteristics. Most of these were constructed in bright colors or in black and white for further embellishment by the consumer.

Color Trends and Selection for Product Design.
DOI: http://dx.doi.org/10.1016/B978-0-323-39395-9.00007-4

In the art world, the artist, Yves Klein deposited a Soleau Envelope in 1962 for his bright lapis lazuli blue color, aptly named IKB, commonly known as Yves Klein Blue or International Klein Blue. The purpose for this registration was to protect the purity of his color that was created from ultramarine pigment suspended in Rhodopas, a polyvinyl acetate, which give the color its unique hue. Although Klein introduced this and other bright colors in monochromatic canvases (including orange, pink, yellow and red) in the late 1950s the work was poorly perceived and largely misunderstood that were probably ahead of their time. However, this IKB color became an important color for the 1960s and early 1970s eras. Two important Art Movements in the 1960s were Op Art and Artre Povera. Op Art or Optical Art was officially born in 1964 when Time Magazine described this new Art Movement as art that appears to be moving or breathing due its precise placement of lines and curves. Op Art became wildly popular and was soon in evidence in fashion, advertising and interior designs. This movement started as an achromatic style in black and white and a few years later evolved into a polychromatic style. Two artists that were key to this movement were Victor Vasarely and Bridget Riley. Victor Vasarely is a Hungarian-born artist who is a pioneer in this 3D, geometric style of art. His early work was a series of pulsating black and white pieces and emerged into colored pieces starting with his Plastic Alphabet that was painted in shades of red, blue, purple and orange. Bridget Riley was a British painter and designer whose Op Art pieces were considered to surpass those of Victor Vasarely. As with Vasarely, her work started in black and white but evolved into color. In 1967, she created work in the ancient Egyptian Color Pallet and later did a study on the psychological effects of monochromatic pieces, one of her major projects was at the Royal Liverpool Hospital in 1983 where she painted the walls with a series of pinks, blues and yellows that was reported to reduce the application of unwanted graffiti.

The other pivotal Art Movement of the 1960s was Arte Povera or "Poor Art." This movement originated in Italy, evolving from the decline of Abstract Art and is characterized by its use of everyday materials such as burlap bags, newspapers and bricks. Italian natives Mario Merz and Luciano Fabro were two of the key artists. One of Mario Merz most famous pieces was his Igloo series that were created from found objects and consisted of materials such as metal, glass, stone, mesh fabric, etc. that were electrified and illuminated with neon and incandescent lights that were positioned to shine through the dirtiness or to send a message using neon signs. In 1996, Merz developed a fashion show with Jill Sander where he created a wind tunnel containing white cloth that was twisted with leaves and sticks.

Luciano Fabro was the unofficial father of the Arte Povera movement and was influenced by artists such as Yves Klein and Lucio Fontana. One of Fabro's most famous sculptures was composed of a mirror with part of the coating scraped off. This piece was entitled "Il Buco" or "The Hole" and is intended to represent the dichotomy of ones identity with the emptiness of society. Fabro frequently worked with such raw materials as rolls of tightly wound copper wire, strips or curls of paper, tubular metal and molten glass.

In 1963, Josef Albers published his ground-breaking book, *Interaction of Color*. This work was a fresh look at the basics of color optics and explored color in context where colors, when layered or appear next to each other influence their hue. As designers and color chemists work with developing new colors or pallets for products, this is an important area of research in order to see how one color component may impact the space that surrounds them rather than looking at distilled hues that are separate from their immediate surroundings.

In the mid-1960s, the Beatles also influenced fashion trends with their psychedelic floral shirts, leather sandals and paisley suits in pastel colors of yellows, blues and greens. In 1968, The Beatles released their "Beatles" album that was commonly referred to as the White Album and John Lennon celebrated the minimalist fashion with his signature white suit.

The 1960s nail and lip colors were slightly pink, and sometimes completely white. There was a sense that vitality was being erased or obliterated and could be linked to the wide-spread use of illicit drugs that dominated the 1960s. The popularity of white lipstick was associated with girl rock bands such as the Shirelles. The fashion designer, Emilio Pucci, was also a prominent influence in the 1950s and the 1960s. His signature prints were psychedelic in pattern and ranged in colors from acid pinks and greens to lavender and magenta and were influenced by Roy Lichtenstein and Andy Warhol.

The appliance colors that dominated the 1960s were primarily earth tones. Colors such as Avocado Green and Harvest Gold were key to this time period and Cinnamon Red was somewhat less popular. Again, these were colors that became popular in the automotive sector in the 1970s. In 1963, Kenner/Hasbro introduced their first "Easy Bake" toy oven, in colors of turquoise, yellow and orange. The history of the Easy Bake Oven is quite interesting and mirrors the color and social trends reflected in the larger society. In 1969, the first Avocado

Green version of the Easy Bake Oven was produced. It also was designed with a hood and a nonworking clock.

In 1962, Crayola changed the name of their neutral color crayon named, "flesh" to "peach" to combat racial insensitivities.

1962 was also marked the birth of the New York Mets Baseball team. The iconic blue and orange colors were chosen to indicate the genesis of the team from the blue Brooklyn Dodgers and the Orange New York Giants that had left the New York area. This color scheme was also used in the 1964–65 NYC World's Fair to gird its exterior panels a short distance from the Mets Stadium in Queens, NY. The US Space Park at the World's Fair featured dancers in neon jumpsuits and atomic hats that attempted to explain space-age science to the casual observer and ignite the world's passion about the emerging space exploration program.

Googie architecture emerged in the late 1950s and peaked in 1964 at the World's Fair. This space-age architecture was an exaggerated modern style that was heavily influenced by the post-World War II automotive culture from the 1950s with sweeping facades and bold colors. This architectural style was heavily criticized as a fad but it now currently recognized a true architectural movement. Few examples of this style still exist today in the United States, most are in Southwest and many of these still function as diners or bowling alleys and are reminiscent of the Jetson cartoon that was popular with children in the early 1960s. One of the Jetson's family pets was Orbitty who expressed his emotions by changing colors. Judy Jetson, the teenage character of the cartoon, embodies the fashion of the 1960s with her platinum-white hair and fuchsia space-age clothing that has a Googie feel of geometric shapes and tidy, ultra-tight lines. Googie color schemes were dominated by turquoise, corals, pinks, light yellows and light blues.

In the 1960s, the American automotive manufacturers were losing market share to international imports. The Big Three (General Motors, Ford and Chrysler) had over 90% of the American market and about 50% of the global market. However, to combat the global threat, the Big Three decided to segment their nonluxury vehicles into three broad classes; economy cars, muscle cars and pony cars. The 1960s vehicles focused on performance and each of the Big Three manufacturers tried to produce the car with the largest engines possible to out-do their competitors. The muscle and pony cars were painted with less traditional, higher visibility colors like yellow, orange and violet while the more traditional sedans and family cars remained firmly entrenched with the 1950s colors. As fins and fenders became dimished, so did the two-toned and three-toned paints also decreased in the 1960s for the same reasons. The exuberant colors were in part due to the use of heavy metal-based pigments such as molybdate orange and cadmium yellow, lead carbonates and lead chromates. These pigments were later banned for use in automotive paints due to their toxicity and these vibrant shades have never been replicated. In 1966, the Bertone-styled Lamborghini in stunning shades of Lime, Electric Blue and a Raspberry Red. The 1960s were also known as the decade when the unfortunate trend of vinyl or laundau tops appeared as faux convertible tops on the roof of vehicles; mostly in black but also seen in various shades of tan, white and green. This trend started in the United States and then spread to Britain. Eventually, Citroen, Audi, Porsche and Opel also picked up this trend a few years later. Another trend that became firmly entrenched in the 1960s was the use of metallescent and pearlescent paints. These metallic and metallescent trends continued to be popular for the next 30 years.

Further Reading

https://en.wikipedia.org/wiki/1960s_in_fashion
https://en.wikipedia.org/wiki/1964_New_York_World%27s_Fair
https://en.wikipedia.org/wiki/Googie_architecture
http://www.spaceagecity.com/googie/
https://books.google.co.uk/books?id=GX-uAAAAQBAJ&pg=PA130&hl=en#v=onepage&q&f=true 1960's
http://www.hasbro.com/en-us/brands/easybake
http://www.pigments.com/Products/Inorganic-Pigments/Chrome-Yellow-and-Moly-Orange/NM-33-FM-Molybdate-Orange-Y-S.aspx
https://en.wikipedia.org/wiki/List_of_The_Jetsons_characters
https://en.wikipedia.org/wiki/Mary_Quant
http://www.thestyleking.com/compass/the-beatles-mens-fashio/
http://laughingsquid.com/the-souper-dress-a-warhol-inspired-pop-art-paper-dress-by-campbells/
http://www.enjoy-your-style.com/white-lipstick.html
https://en.wikipedia.org/wiki/Yves_Klein
http://www.wikiart.org/en/eva-hesse/expanded-expansion-1969#supersized-artistPaintings-302502
http://www.theartstory.org/artist-bourgeois-louise.htm
http://www.britannica.com/biography/Luciano-Fabro
http://www.nytimes.com/2007/07/03/arts/03fabro.html?_r = 0
https://en.wikipedia.org/wiki/Mario_Merz#Work
http://www.tate.org.uk/art/artworks/merz-igloo-do-we-go-around-houses-or-do-houses-go-around-us-t05755/text-summary

8 1970s Trends: The "Me Decade"

The 1970s were also known as the "Me Decade" in the United States. This was a time that young people focused on themselves rather than the world at large and is thought to be related to the prosperous post-war time era along with disappearance of the working class and the emergence of a prominent lower class of society. Prior to the 1970s, fashion tended to be divided into two categories: high-end couture and ready-to-wear (RTW) mass-market clothing. The 1970s marked the beginning of designer brands. Fashion designers and hairstylists joined the A-list of celebrities that were known throughout the world and across economic sectors. It also marked the difference between Mass-Market and Ready-to-Wear clothing lines. This was an important distinction because it made, "Designer Labels" accessible to the average consumer and further segmented the garment industry into three broad categories: haute couture, RTW and mass-market. This demarcation of the RTW/mass-market categories was also strengthened by the shift in production from European and American factories to Asia and later to Latin America that favored cheaper labor costs and lax regulatory standards reminiscent of American and European sweatshops in the late 19th and early 20th centuries. Many of the American factories bundled up their sewing machines and other equipment and shipped them off to Asia for rapid development. However, Asia was also quick to invest in and develop new types of high volume and computerized methods of manufacturing that has kept their place in the market unrivaled to this day. As consumers and designers, increased their global travel, fashions of the 1970s also were quite diverse and had a wide range of expression from tailored disco, to voluminous caftans, from hippy naturals to vibrant synthetics. In 1974, Diane Von Furstenberg launched her signature wrap cotton/rayon dress in a green and white dotted pattern. Hippie styles continued to flourish from the 1960s and featured influences from Indian and African cultures that embodied earth-toned natural fibers, tie-dyed fabrics decorated with flowers and fluorescent peace signs were reflective of society's disenchantment with inauthentic militaristic cultures of western civilizations. In 1970, Lee Jeans commercialized their casual suit that consisted of a shirt jacket and

Color Trends and Selection for Product Design.
DOI: http://dx.doi.org/10.1016/B978-0-323-39395-9.00008-6

matching pants comprised of synthetic fabrics in solid bright colors like orange, green and wine colors or in plaid constructions of the same colors. Lee Jeans dubbed it the "Lee-Sure" suit also known as the Leisure Suit. Fortunately, the Leisure Suit declined in fashion in the early 1980s but there are a few couture designers that revive the style every few seasons. It is important to note that this casual suit is not a new development. It is possible that the leisure suit evolved from the British Safari Jacket around the turn of the 20th century. There was also a Hollywood Suit in the 1950s that were worn by performers and frequently adorned with rhinestones and other decorative motifs. What made the 1970s leisure suit unique was its popularity with the middle class, the use of synthetic fabrics as well as the vibrant colors that were worn by both men and women. Disco fashions were very influenced by the heavy beat of disco dance music and the club life that embodied this culture. Tight-fitting, synthetic, elastomeric clothing in vibrant shades of red, tangerine, mint green, white and gold lamé are reminiscent of the John Travolta, Saturday Night Fever, iconic figure. Although clothing was extremely bright in the beginning of the 1970s, by the end of the 1970s the colors of the clothing had faded to shades of gray, white and blacks and earth tones due to consumer color fatigue.

The 1970s was also an era of gender-neutral (antifashion) clothing became firmly entrenched. Although the trend started in the 1960s with the Women's movement, the antifashion, antifeminine styles became so mainstream that the Sears catalog removed all pink toddler clothing from its pages for 2 years. Popular colors in this antifashion movement were muted primaries and neutrals, layered with earth tones. Popular clothing styles included button-down flannel plaid shirts and blue denim and corduroy pants for both sexes. This was also the time when punk rockers emerged in their plaid pants and angry reds contrasted with tough black leather motorcycle jackets and working-class boots indicative of the hardships faced by the declining middle class in the United States and Western Europe.

For similar reasons, appliance colors continued to fade in the 1970s and tended to mirror food colors that were more natural such as Almond, Wheat and Coffee. Avocado and Harvest Gold continued to dominate in the kitchen in 1970s but a new bright Poppy Red also emerged. As the 1970s continued, fresher, more natural kitchen colors resurfaced such as Espresso and a cleaner version of Avocado. Turquoise and Copper continued to be popular in kitchens throughout the 1970s. In 1970, Hasbro introduced their Harvest Gold Super Easy Bake Oven that was twice the size of the earlier version. It also included a functional timer.

In 1978, they introduced a white, orange and brown version of the Easy Bake called the Mini Wave. It also included a nonfunctional digital clock.

Lip and nail colors of the 1970s were whimsical; almost childish; bright blue-shade purples, frosted lime green and dark blue with silver flakes dominated the era. This could be attributed to the emphasis on fashion and the coordination of makeup and clothing on the runway. Black lipstick also came back in fashion and remained into the 1990s with the development of Punk and Goth cultures.

As a backlash to the oil embargo of 1973 by OPEC and the gas-guzzling muscle and pony cars of the 1960s, consumers were looking for greener and more fuel efficient cars in the 1970s. Governments became more involved in air quality standards relating to automotive manufacturing, including painting and exhaust emissions. As a result, cars became smaller and the color schemes were less performance-oriented; bright reds, yellows and oranges were replaced with more organic; browns, tans, creams, mossy green and earth tones. The AMC Pacer was an American car that truly embodied the automotive spirit of the 1970s. This was a car that was designed from the inside out and was the widest, smallest car on the market meant to enhance the rider's experience through additional passenger room and an excess of rounded glass. The Pacer's exterior colors ranged from "Mellow Yellow" to "Loden Green." AMC also paid attention to the interior, offering a Levis Blue Jean Interior or a Navajo Design seating area with wood grain design on the instrument and interior door panels. The Pacer was discontinued in 1979 due to poor braking and handling, lack of storage space and a high curb weight that contributed, in part, to its poor performance. Other detractors felt that the car's bubble design was esthetically unappealing. AMC also tried to export the Pacer into Europe and Mexico and was unsuccessful for similar reasons in Europe but also due to the higher price point in Mexico. Bahama Yellow was an enduring automotive color that emerged in the 1960s and endured through most of the 1970s. Porsche also introduced a line of performance colors including Signal Green and Yellow and Orange that gave these high-speed cars great visibility under the auspices of safety since they were visible in almost every weather condition.

Both feminist theory and the feminist movement were great influencers of the Feminist Art Movement that started in the 1970s in the United States. Two primary artists of the Feminist Art Movement were Eva Hesse and Louise Nevelson. This art movement was seen as a shift from traditional Modern Art and demanded parity with the highly entrenched, male-dominated art world. Although the trend started in the United States, it quickly spread to Europe, Australia, Latin America and Canada. Some of the themes that were addressed were women's bodies, domestic life and emotional experiences. Eva Hesse was German-American artist. She was a German Jew that was exiled by Kindertransport to Britain and eventually to New York during World War II. Her work utilized modern materials such as plastics, fiberglass and latex. She later returned to Germany with her husband but found it difficult to work and express herself creatively, returning to New York in the 1960s. One of Hesse's most celebrated sculptures is her, "Ringaround Arosie" which is a sculpture of a breast and a penis that has a stereotypical feminist perspective with its pink tonality and craft-like modality. The name of the sculpture is derived from the children's game where children fall down as they hold hands and run in a circle, playing out the details of the Bubonic Plague or Black Death that occurred in London in 1665.

Louise Nevelson was best known for her monumental abstract sculptures that were developed at a time when mostly men created large, imposing structures in metals and stone. Louise's work was primarily constructed in wood and one of her most famous pieces was "Sky Cathedral" which was a large collection of black boxes created from scrap wood that were nailed together. The black color is an important component to this work because it represented the collection of all colors together as a totality. The black color was also used to harmonize the disparate shapes and negate their past.

Further Reading

http://paintref.com/cgi-bin/colorcodedisplay.cgi?manuf=AMC&model=Pacer
https://en.wikipedia.org/wiki/AMC_Pacer
http://www.enjoy-your-style.com/black-lipstick.html 70/80
http://www.metmuseum.org/toah/works-of-art/1997.487 (70s)
http://paintref.com/paintref/model/green_Pacer.shtml
https://en.wikipedia.org/wiki/The_%22Me%22_Decade_and_the_Third_Great_Awakening
http://sammydvintage.com/vintage-style/70s/1970s-fashion/
http://www.obieawards.com/events/1970s/
http://www.fashionencyclopedia.com/fashion_costume_culture/Modern-World-Part-II-1961-1979/Leisure-Suits.html
https://en.wikipedia.org/wiki/1973_oil_crisis
https://en.wikipedia.org/wiki/Feminist_art_movement

9 1980s Trends: Conspicuous Consumption

The 1980s were explosive with houseware brands that were both high-end and disposable. Appliance colors tended to be stark choices of black and white. This surge in high-tech/low-color appliances might be due, in part, to men sharing in cooking and housework that started in the late 1970s and was characterized by dominance of metal fixtures and colors contrasted with plastic furniture in bright primaries 1980s kitchen design was characterized by light wood tones, combined with slick black and white surfaces. Countertops and floors were covered with thick, glazed earthenware tiles. The kitchen island also became very popular in the 1980s. The American country look of the early 1970s was replaced by the French Provincial look in the 1980s with the emergence of more blue accent colors and less red and green shades in the kitchen. Another trend that became prominent in kitchens was the design of built-in seating and banquettes providing ample room for guests and entertaining in the kitchen. Many of these were created in sinuous shapes using Memphis styles and colors of upholstery.

In 1985, Ikea opened their first furniture store in the United States. Although Ikea was founded in the 1940s in Sweden, it gradually migrated to the rest of Europe, Asia and Canada in the 1970s. The Ikea store is easily identified as a large bright blue box with yellow accents, similar to the national colors of Sweden. Ikea has a major influence on middle-class domestic design with their emphasis on Scandinavian modern or mid-century modern styles that are typically created with light, natural wood tones that are punctuated with bold and earth-toned shades of oranges, blues and greens.

The Easy Bake Oven was redesigned in white and featured a high and low temperature setting. In 1982, IBM launched their first personal computer and rapidly found its way into the workplace, replacing mini-frame and mainframe systems. Apple launched its Macintosh computer in 1984 and became a favorite among hobbyists and home users. Commodore launched its Vic64 in 1982 and was the most popular home computer on the market. All of these computers were issued in a

Color Trends and Selection for Product Design.
DOI: http://dx.doi.org/10.1016/B978-0-323-39395-9.00009-8

standard off-white or taupe colors whereas Walkmans, Boom Boxes, VCR players and digital pianos from Asia that were also introduced in the 1980s were marketed in shades of blacks and silvers. Sony also introduced their new Trinitron televisions in the 1980s that exhibited superior color and resolution when compared to their Western counterparts. However, they also changed the style of televisions. In the 1950s through the 1970s, American-manufactured TVs were enormous furniture-like consoles that graced the living rooms of the middle and upper classes. These new high-performance televisions of the 1980s were smaller, in black and silver tones and were sometimes encased in synthetic wood grain laminates. Televisions also became common in third world countries and the number of televisions in India and China increased 10-fold in the 1980s decade. The Internet also was born in the 1980s and by the end of the decade, over 1 million Americans were on-line.

Although bean bag chairs were officially invented in 1969 in Italy, they became heavily commoditized and popular in the 1980s when they were marketed as casual youth furniture in embossed PVC in a variety of colors ranging from Federal Blue, Fiesta Red, Strawberry, Autumn Spice, Ebony and Jungle Green.

Lip and nail colors of the 1980s were powerful dark burgundy or claret colors. This could be related to the rise of women in the workplace seeking parity in compensation and status.

This was also when the rise of the Armani power suit with large shoulder pads became popular for the same reasons. Power dressing was important for both genders in the 1980s. Since women were seeking and sometimes gaining financial parity with men in the same positions, they invested in more expensive and serious clothing in conservative shades of navy, gray and black, frequently paired with a brighter bow blouse in shades of pink, violet or mint. Men also wore these power suits in the 1980s that were characterized by narrow lapels, broad shoulders, wide pinstripes in shades of blue and gray.

Fashions and colors changed rapidly during the 1980s. Colors in the early 1980s were a carry-over from the 1970s and were quite subdued shades of tan, brown and burnt orange. Japanese minimal fashions were highly textured, in numerous shades of black, charcoals and dark browns. The origins of the Japanese fashions and colors are sometimes ascribed to the emerging homeless culture of those unfortunate members of society whose social programs were cut out of government funding, leaving the less fortunate to fend for themselves on the streets after being evicted from hospitals, half-way houses and government subsidized housing in the United States and Europe.

In 1989, the Berlin Wall was torn down in Germany and this marked a dramatic cultural shift not only in Germany but the rest of Europe as well. This signaled the decline of Communism as many of the resource-starved, Eastern European countries embraced consumerism, looking to the west for new fashion, technology and color trends.

However, the colors heated up as the decade progressed. Consumers grew tired of the bland and dirty hippie earth tones of earlier decades and wanted to add some colors. There were five subcultures that influenced fashion in the 1980s: New Wave, Rockabilly, Hip Hop, Preppy and Glam Metal. The New Wave subculture was in direct opposition of the Hippie subculture and evolved from the Punk/Goth cultures of the 1960s and 1970s. New Wave colors were shades of black jackets, trimmed in gold and white pirate shirts combined with dark leathers. The Rockabilly subculture was linked with the Teddy Boy Styles of the 1960s and 1970s in the United Kingdom. However, the suit jackets were substantially darker in shades of deep blue, dark gray and darker maroons than were popular in earlier decades. White t-shirts, baseball and black leather jackets were also popular in this subculture. The Hip Hop subculture was a logical extension of the disco dance era of the 1970s because of the club-like connections between the two subcultures. High-end athletic shoes were popularized by this subculture and similar to the Teddy Boys' Creeper shoes, they were perfect for high energy dance moves. Red, Yellow and Green (colors of Africa) became essential shades of Hip Hop styles. Pange and Kente Cloth colors (orange, blue, red) and styles were also a part of the Hip Hop style of dress. The Preppy subculture was a very American movement that was characterized by very traditional clothing in a variety of khaki colors and fabrics, combined with brightly colored madras and polo shirts including pink and bright greens for men and women. Glam Metal was a musical subculture that evident in the United States, Europe, the United Kingdom and Japan. Fashion colors included acid-washed, blue and black denim, black leather or satin jackets, dark brown and black animal-printed shirts and jackets. Glam Metal was replaced by Grunge in the 1990s.

Brightly colored aerobics clothing for women and velour track suits for men in shades of brown, orange, black and green were also popular as casual clothes in the 1980s.

Princess Diana was a fashion and style icon of the 1980s. Prior to her marriage, her clothing styles were quite simple country, flowered dresses and skirts but once she became the Princess of Wales, she was pushed by the fashion industry to dress in a more sophisticated manner. In a few years, she defined her own style and became an international icon. Her embodiment of millinery styles breathed new life into this ailing industry and she was frequently photographed in vibrant shades of red, jewel tones and soft pastel tones.

In 1984, Hermes created their first Birkin bag for fashion icon Jane Birkin in a black embossed satchel. Current colors include Clemence Red, Bamboo Green, Taxi Yellow 30, Hot Feu Fire Orange, Rose Tyrien, Aubergine Purple, Blue Electric, Classic Black and a variety of hides including alligator, ostrich and lizard.

Prenatal testing allowed new parents to determine the gender of their unborn child prior to birth so gender-specific clothing, furniture, accessories and supplies in shades beyond yellow and white became an emerging market.

1980s automotive colors became more traditional, replacing the earth tones and socially responsive colors of the 1970s with more conservative colors like black and gray. Black and bright red were the most popular automotive colors in the US automotive market. It is important to mention that automotive color trends vary slightly throughout the world, however in the 1980s, the Big Three automotive companies (Ford, Chrysler and General Motors) were still based in the United States and dominated the global market in terms of style, design and market penetration thus automotive color trends that started with these three companies had a large influence throughout the rest of the world. Obviously this changed in later decades when imported vehicles from Asia and Europe penetrated and dominated the American market. Their designation was changed from "The Big Three" to "The Detroit Three." Another distinctively American style trend that was still prominent in the 1980s was the use of faux wooden PVC decals that were adhered in a two-tone arrangement to the exterior of cars. The emergence of the Chrysler minivans in 1984 became a popular replacement for the station wagons that were a staple family car in the earlier decades. Two prominent colors for the minivan was a Black Cherry and Metallic Blue. The vehicles were also offered in a metallic silver and gold in later years. In 1988, Mary Quant (mini skirt and hot pants designer) designed the interior of the BMC Mini. Seating surfaces were black and white striped trimmed in red. On the exterior, the headlight surrounds and bumpers were matte gray while the body colors were either diamond white or jet black.

Three important Art Movements of the 1980s were Fractalism, Animation and Graffiti. Fractal Art is a type of computer art that is developed from mathematical algorithms. The invention of the silicon chip in the 1970s and the Mandelbrot Set, also known as "The thumbprint of God," in the 1980s, created a revolution in computers and allowed for the rise of Fractal Art. The Mandelbrot Set is a series of computer-generated images, infinite in color and detail. The colors are arbitrarily assigned and are called Cycled Colors that start at the zero point of black and reach out to infinity using infinite hues and combinations of colors.

Graffiti was not a new art form that started in the 1980s but rather can be traced back as early as the cave paintings in Lascaux France. As the art world became more exclusive and the majority of artists were shut out of the opportunity to share their art, Graffiti became a medium for taking art back to the people and as a mode for self-expression. The most striking part of graffiti art is its relationship to urban cultures, gang-related violence and vandalism as well as a disenfranchised youth culture. For these reasons, graffiti was commonly viewed as vandalism rather than art and it was not until the 1980s when the art community saw the decline in graffiti decided to bring it in off the streets and into the galleries. Part of the reason for the decline of street art was twofold. The first was the development of antigraffiti surfaces and paints. Graffiti removal is a big business globally. In 2004, Germany spent over 400 million USD to clean up unwanted graffiti. The United Kingdom spent over 1.5 billion USD and Los Angeles California spent over 140 million USD in just the inner city. There are two basic types of antigraffiti coatings: sacrificial and permanent. The sacrificial coatings absorb the graffiti and then are washed off or sacrificed with high-pressure water. The surface is left clean and the sacrificial coating must be replaced to absorb the next application of graffiti. The permanent coatings have many different chemistries including nano-particles and fluoropolymers that will prevent graffiti from permanently adhering to the surface. In most instances, graffiti can be removed by scrubbing the surface with solvents but the antigraffiti coating remains unphased. The other reason for decline of graffiti in public is due to the use of high-tech surveillance such as increased illumination and security cameras. Two of the most celebrated street or graffiti artists are Banksy and Jean Michel Basquiat. The anonymous British, Bansky, is more antiprofile than anonymous. His fame for his anonymity is almost as intriguing as his satirical artwork. Although Banksy's work has grown more prominent in the 21st century, it is believed that he started his graffiti career in Bristol in the 1980s when he began using stencils as a method of creating high-quality paintings in a short period of time and avoiding arrest for vandalism. Most of his work is in black and white with pops of bright colors. Jean Michel Basquiat was born and raised in New York City by Hatian and Puerto Rican, immigrant parents. Although his career started in the late 1970s, he continued to gain prominence with his provocative graffiti statements about race and class with his Graffiti group, SAMO that covered the buildings of New York's lower East Side with cryptic phrases. In 1979, he published his announcements on the walls of SoHo buildings that "SAMO is Dead." His fame grew in the 1980s with his associations with Andy Warhol and Pop star Madonna. His dark art started to show in New York galleries and attempted to demonstrate the plight of being a black man in America. His work contained vestiges of his earlier graffiti, containing graphic text and images that used piercing contrasts of bright color that were a hallmark of his personal brand.

Animation was another important Art Movement in the 1980s. Two prominent artists of the 1980s that were identified with animation were Keith Haring and Osamu Tezuka. Osamu Tezku is sometimes referred to the "God of Manga" and his work is Robot or Super Robot drawings, illustrations, comics and toys. One of the most famous is the UFO Robot Grendizer, the DVD cover featuring bright primaries of reds, blues and yellows combined with futurstic line dawings of UFO's and space ships. *Mangaka* is the Japanese word for comic artist. Manga is generally referred to illustrations in a comic book whereas mangaka refers to the artist of the manga. The word can be broken down into two parts: manga and ka. The ka suffix refers to an official degree of expertise or skill in drawing and the manga is the art itself.

Keith Haring was a New York City artist who was famous for his bold, brightly colored drawings that were frequently politically charged. Haring also was well known for his graffiti works in his early career and his heavy-lined images are frequently associated with such artists as Andy Warhol or Roy Lichtenstein.

Further Reading

http://www.allpar.com/model/m/history.html
https://en.wikipedia.org/wiki/1980s
https://en.wikipedia.org/wiki/1980s_in_fashion
http://www.housebeautiful.com/room-decorating/kitchens/g409/kitchens-1980s-0609/?
https://en.wikipedia.org/wiki/IKEA
https://www.allpar.com/model/m/original-minivans.html

https://en.wikipedia.org/wiki/Plymouth_Voyager
http://www.pujesdiary.com/2013/12/10/jane-birkin-and-history-of-the-birkin-bag/
https://en.wikipedia.org/wiki/Berlin_Wall
http://www.dailymail.co.uk/femail/article-2116038/The-Kate-effect-strikes-Sales-coral-coloured-jeans-soar-Duchess-wore-pair-play-hockey.html
https://en.wikipedia.org/wiki/Anime
http://www.mnn.com/lifestyle/arts-culture/stories/global-graffiti-8-powerful-street-artists
https://en.wikipedia.org/wiki/Jean-Michel_Basquiat
https://en.wikipedia.org/wiki/Anti-graffiti_coating
http://artisticmovementsofthe1980s.blogspot.com/2008/11/graffiti.html
http://www.banksy.co.uk/index5.asp
https://en.wikipedia.org/wiki/Grendizer
http://www.basquiat.com/artist.htm
http://topdocumentaryfilms.com/fractals-colors-infinity/

10 1990s Trends: Electronics

Personal computers, mobile phones and the Internet changed the way we gathered information, learned and transmitted ideas. In the 1980s, it was predicted that cities would die and its residents would move far away from urban centers. This was due, in part, to the birth of the World Wide Web in 1990 that was erroneously predicted to rapidly reach every part of the globe and that people would not need to be close to each other but could access the larger society through these electronic connections. What happened was just the opposite. In emerging countries in Asia and Latin America, rural populations moved to the urban centers to improve their quality of life. In Europe and the United States, we saw that cities continued to thrive, suburban areas tended towards modest growth and rural cultures tended to stagnate.

Kitchens were no longer the only rooms that tended to need appliances of color. In 1994, HGTV (Home and Garden TV) was launched onto the airways and revolutionized how many television viewers decorated their homes. In 1999, the same production company launched the DIY network. These infotainment shows, empowered home owners to make bolder color choices in their wall, floor, furniture and surface finishes in the 2000s. Interior colors included Navy, Taupe, Rust and Putty. This will shift to brighter colors in the 2000s. This is another example of a cyclical trend in our society.

There were also new series of high-end major appliances such as washers and dryers as well as vacuum cleaners that were available in choices other than white such as reds, silvers, blacks, etc. An interesting migration of appliances in the home also occurred in the 1990s; refrigerators that had once been relegated to the corner of a kitchen or a pantry became the heart of the kitchen, swelling in size, and bursting with additional functionalities including liquid and ice dispensers as well as information centers with digital displays that indicated internal temperatures, freshness dates of the contents and built-in entertainment systems. The same was true of laundry appliances that were previously stowed in a laundry room or a basement that shifted into attached garages or into kitchens or hallways and became much more visible. In 1993, Hasbro reintroduced

Color Trends and Selection for Product Design.
DOI: http://dx.doi.org/10.1016/B978-0-323-39395-9.00010-4

their "Easy Bake Oven & Snack Center" in a gender-specific pink that contained a warming tray on top of the oven to melt toppings.

Lip and nail colors of the early 1990s were more natural and understated shade of browns and quite matte in finish. The later 1990s favored more glossy finishes with heavy concentrations of luster pigments.

In 1992, Crayola offered their new Multicultural Line of eight crayons that denoted a more realistic set of skin hues from Apricot to Burnt Sienna, including black and white for blending. In 1999, they renamed their "Indian Red" crayon to a more politically correct "Chestnut" so as not to offend American Natives even though the pigment used in the crayon is a pigment that originated in India.

The 1990s automotive colors shifted away from the primary and flat colors of the earlier decades and offered a fresher pallet of fusion colors such as blue-green aquas, red-blue burgundies and softened metallic colors. Two-toned color paint schemes came back on trend and were celebrated with the birth of Sport Utility Vehicles. The most popular automotive color of the 1990s was green. The 1990s was marked as by international invasion of automotive manufacturers in the United States. Volkswagen Mercedes and BMW opened SUV manufacturing facilities in the southern part of the United States. Toyota, Honda, Hyundai and Kia also began manufacturing their vehicles in the US.

As ostentatious as the fashions were of the 1980s, the fashions of the 1990s were equally minimal and casual. The shoulder pads, large gold jewelry and teased hair were replaced with unstructured jackets and cargo pants. In 1997, Hong Kong was returned to China from the British and the media attention of this event influenced the fashion world to incorporate touches of Chinese traditional styles, colors and finishes including jewel-toned silks and satins in shades of crimson, emerald, jade and sapphire. Rich, beaded and brocaded jackets, hand bags and dresses were the perfect accessories for the pared down fashions of the 1990s. High-waisted pants and jeans became popular for men and women. Kate Moss and Madonna were two pivotal icons in the 1990s. Kate Moss was known for her emaciated "Heroin Chic" image with her sunken eyes and bloodless skin tone. She was also known as the antimodel of her time and frequently appeared in dark and dirty grunge fashions. In contrast, Madonna, "The Material Girl," was much more buxom and colorful and was frequently photographed in layered lingerie in vivacious shades of pink, orange, gold and every shade of red. In 1992, Christian Louboutin introduced his stiletto high heeled shoe with a vibrant red outersole. According to Louboutin, the color was inspired by his frequent visits to a museum in Paris that had a sign with a stiletto that was covered by a big red X to indicate that these shoes were forbidden inside the building due to the damage they could cause to the wooden floors.

According to the New York Times, 1992 was "the year of the Ribbon." This is in reference to awareness ribbons that people wear to express their support for a certain cause or their disenfranchisement from a specific situation. There are ribbons in almost every color of the rainbow with infinite causes. As an example, a green ribbon has at least 15 causes it may represent public school support, Lyme Disease, pedestrian safety, Mitochondrial Disease, environmental protection, Cerebral Palsy, etc. The display of ribbons or flowers was not a new idea. In the United States, the red paper poppy, also known as the Remembrance Poppy has been displayed since 1921 to commemorate all of the soldiers who were killed in wartime struggles. It was inspired by the "Flanders Field" poem written by John McCrae who wrote how Poppies were the first flower to grow around the multitude of graves in Flanders Field. These Poppies were also used to recruit soldiers and promote the sales of war bonds. The French had their own version of the red poppy called the Bleuet de France. This was named after the blue cornflower that became the symbol of the young French soldiers that fought and died in World War I that wore Horizon Blue uniforms.

In 1992, the Toyism Art Movement was founded in the Netherlands. The defining characteristics are the use of bold and bright colors in paintings, sculptures and structures. The use of lines, dots and shapes defines the energetic and playful nature of the global movement. In 1994, Jeff Koons, American Pop artist, launched his Celebration series of sculptures that consisted of Easter eggs, diamonds, hearts and balloon dogs. These larger-than-life sculptures embody the fin de siecle opulence that was associated with the turn of the century Art Movements. The Celebration series was intended to signify the ecstatic experiences of childhood holidays including birthdays, Easter and Valentine's Day. The larger-than-life dogs stand over 10 feet tall but the perfection of its form suggests the weightlessness of an inflated balloon. Although the balloon dogs had the appearance of a latex balloon, they were actually fabricated from stainless steel and painted with a candy-coat of either orange, magenta, blue, red or yellow dyes. The orange dog sold at Sotheby's in 2013 for 58.5 million USD.

Further Reading

http://web.archive.org/web/20140415055039/http://www.npr.org/2011/10/16/141402115/breast-cancer-when-awareness-simply-isnt-enough
https://fr.wikipedia.org/wiki/Bleu_horizon
https://en.wikipedia.org/wiki/Bleuet_de_France
https://en.wikipedia.org/wiki/In_Flanders_Fields
https://en.wikipedia.org/wiki/Remembrance_poppy
https://en.wikipedia.org/wiki/DIY_Network
https://en.wikipedia.org/wiki/Lolita_fashion
http://www.christies.com/lotfinder/sculptures-statues-figures/jeff-koons-balloon-dog-5739099-details.aspx#top
https://en.wikipedia.org/wiki/Jeff_Koons
http://toyism.com/home
http://www.visual-arts-cork.com/postmodernism.htm

11 2000–2006: September 11

The 2000s started with the trend of space-age or futuristic cosmetics associated with the new millennium. Silver, gold and copper metallics permeated face powders, blushes, bronzers and nail enamels. More pink shades were evident in the early 2000s than in the 1990s but still quite understated. These formulations also became more glossy and the use of pearlescent and color-shifting pigments became wide-spread this might be, in part, due to the economic prosperity of this era.

Y2K hysteria fueled a series of off-the-grid products and technologies. In general, the larger society was concerned about the collapse of computer-driven systems of everything from Wall Street and utility companies to water reservoirs, resulting in financial and natural disasters. New self-sufficient products emerged on the market such as hand-cranked radios and flashlights, survival food products and emergency temporary shelters and transportation devices. Most of these products were styled in military, survival colors of khaki, olive, earth tones dark shades of black and brown and, of course, Red Cross red.

After September 11, lip and nail colors of the 2000s were also following this bruised appearance with shades of brown and purple. At this time, Gerald Lauder of Estee Lauder was credited for coining the phrase, "Lipstick Effect" since his company saw an increase in lipstick sales after September 11. This is not a new phenomenon and was certainly evident during the World Wars and the depression era where women purchase small luxuries like lipsticks or nail products to boost their mood and increase their attractiveness to more financially viable partners who are also in shorter supply during times of economic hardship.

The four simultaneous terror attacks in the United States on September 11, 2001 impacted all of the global trends and color trends were no exception. Prior to this, color pallets were heavily influenced by European industrial activities including automotive, electronics and architectural. These colors tended to be earthy and oxidized contrasted with bright blues and oranges. After September 11, fashion colors became brighter and more explosive. Consumers needed an escape from the para-militaristic olive drabs and khakis and needed an outlet for

Color Trends and Selection for Product Design.
DOI: http://dx.doi.org/10.1016/B978-0-323-39395-9.00011-6

cautious optimism and the impending rage and frustration of the second decade of the 21st century.

The Bratz line of dolls were launched in 2001 and were criticized as the anti-Barbie dolls. The dolls were styled in tight-fitting, glittery blue and black denim jeans and revealing halter-neck tops in metallic and bright pink shades. Mattel followed the sexy tween trend by unsuccessfully releasing their "My Scene" dolls a few months later. In 2006, Bratz dolls superceded the sale of Barbie dolls for the first time in history.

This dichotomy between the pageant princess and the street girl closely resembles the fashion trends of the early 2000s that were a mélange of left-over late 1990s styles combined with faux furs and other lines of ethical clothing. There was a direct link between September 11 and the popularity of distressed denim that allowed the consumers to outwardly express the loss and damage that they were experiencing from the global instability that was inflicted from that event. Bruised-colors fashions also became popular for the same reasons. As an example, shapeless, dark print floral dresses with patterns in pink, purple and blue rose in prominence in the United Kingdom. Body modification, including tattooing and piercing, also peaked during this time and also influenced later fashion designs and colors. Ed Hardy's line of Tattoo-inspired clothing rose to prominence during this time. Don Ed Hardy, an American alternative artist whose clothing was characterized by black t-shirts and canvas shoes that were festooned with bright-colored drawings of tattoos accented with rhinestones and metallic accents. This clothing rapidly became popular with less-desirable elements of popular society and was quickly dropped by the mainstream public. Another example of antisocial fashion during this time period was when the British Chav's (Council-Housed & Violent), a disaffected youth subculture, embraced the Burberry plaid and color schemes of red, black and tan. Burberry quickly distanced themselves from this social stigma by removing the iconic plaid from 90% of their clothing and accessories line that has been creeping back into their designs in recent years.

The new millennium also inspired some interesting colors for food targeted to children. In 2000, Heinz introduced their EZ Squirt Blastin' ketchup in green. The green condiment was a hit and sold over 12 million bottles in their first year. In fairness, the success of this product was not due entirely to the color change of the ketchup but rather the redesign of the bottle made it easier for children to grip and dispense the product. In years following, Heinz has introduced ketchup in a few different shades including teal, purple, pink and orange. They also introduced a rainbow mystery color in 2002 where the consumer did not know the color of the ketchup prior to use. According to Heinz, the overall favorite shade remains blue for children's ketchup. In 2001, ConAgra foods introduced Parkay Fun Squeeze, a butter substitute product that was offered in "Shocking Pink" for girls and "Electric Blue" for boys.

The early 2000s was filled with automotive dichotomies. In 2001, Toyota launched their first electric hybrid in the United States: the Prius. The Prius was a remarkable car both in design and spirit. Unlike the typical Japanese cars whose styles closely resembled Western vehicles, the organic, raindrop shape embodied the spirit of the green vehicle movement as well as celebrated the Japanese style esthetic. The Prius became a fashion statement of celebrities and the super-thrifty and everyone in between. It also helped unseat the faltering Big Three automotive manufacturers in the United States: Ford, Chrysler and General Motors. The Prius was offered in a range of six colors including the typical black, white and silver along with the more unusual shades of Light Aqua Metallic Opalescent (sky blue), Electric Green Mica and Blue Metallic Mica. During this same time period, General Motors launched two of their worst vehicles in the early 2000s: the Hummer H2 and the Pontiac Aztec. The Aztec was launched in 2001. This car was one of the first vehicles to be completely digitally designed and a British newspaper dubbed it as the world's ugliest car. This utility vehicle had some useful features such as an optional tent attachment for camping and an all-wheel drive option it only sold 30,000 units in its best year. It was offered in an array of eight different colors: two reds, a black, white and gray as well as Electric Blue Metallic, Fusion Orange Metallic and a mushroom-tinted Sedona Beige Metallic. The Hummer H2, completely out of touch with the green automotive movement, was the larger version of the previous Hummer H1, the first mass-produced military combat vehicle for civilian use, it is the polar opposite of the Prius that was also embraced by the celebrity world with gas mileage in the 8–12 mpg range. The Hummer H2 was launched in 2004 in eight different colors: black white and gray, two reds, Sage Green Metallic, a bright yellow and a rusty orange color. In 2002, Ford relaunched their 11th version of the Thunderbird. This was a beautiful car that ignited the retro renaissance movement in the automotive industry. Other examples of the Retro movement were the Chevrolet HHR (styled after a 1954 panel van) and the Chrysler PT Cruiser (inspired by Al Capone's 1928 Cadillac Sedan getaway car). The Thunderbird was offered in five colors with tone-on-tone matching convertible tops: black, white, red, yellow and their signature Thunderbird Blue which was more of a turquoise shade. In 2005, Bugatti launched its Veyron, the most expensive mass-produced car in the world at 1.6 million USD. It was also the fastest mass-produced vehicles topping out at 256 mph and burns through an entire tank of gas in 12 minutes and a full set of tires in 15 minutes. It was offered in four uninspired exterior colors: black, white, silver/gray and red. In 2003, Volkswagen stopped their production line of the Type 1 Beetle that had been in production for over 65 years. The Final Edition was painted in Aquarius Blue or a Harvest Moon Beige and only 3000 of these vehicles were produced before closing the production line in Mexico.

As mentioned previously, in 1990s HGTV (Home and Garden TV) and the DIY network were launched onto the airways and revolutionized how many television viewers decorated their homes. Home interior colors in the first decade of the 2000s became brighter and fruitier. Lighter colors such as Kiwi, Lime and Persimmon replaced the Taupe, Putty and Rust colors.

Further Reading

http://www.ridelust.com/goodbye-2000s-the-ten-cars-that-defined-the-decade/
http://www.visual-arts-cork.com/definitions/nail-art.htm
https://en.wikipedia.org/wiki/Burberry (2001)
http://www.packworld.com/package-component/closures/parkay-spreads-kids-bottles 2001
http://www.ridelust.com/5-automotive-trends-that-are-thankfully-fading-away/
http://www.cardekho.com/bugatti/bugatti-veyron-colors.html
https://en.wikipedia.org/wiki/Chrysler_PT_Cruiser
http://www.gq.com/gallery/famous-gangster-cars-jamie-kitman
http://usatoday30.usatoday.com/money/industries/food/2003-04-07-blue-ketchup_x.htm

12 2007–2012: The Great Recession

This was the era of angry colors that was due, in part, to the Great Recession that started at the end of 2007, through 2008 and ended in 2009 that was sparked by the bursting of the multi-trillion dollar housing bubble that caused a decline in consumer spending, failing banks and massive losses on Wall Street. Reds dominated the fashion world and pink had faded to the background due to an over-commoditization in Asia. Luxurious neutrals were replaced by earth tones. Moody blacks and grays washed across the color pallets. Similar to the 1960s that started with a reduced and dark pallet that gradually liberated to an enriched pallet of highly chromatized colors, unlike the 1960s, there was an increased use of sparkle, metallescent and color shifting effect pigments. These pigments enriched the pallets of the earlier decades and transformed the cyclical nature of color trends into pluralistic, unique and new retrospective pallets. Colors transformed from moody grays, to transparent grays and eventually transitioned into playful lighter tones of pinks, reds and greens that were clearer and whitened tart tones that were influenced by OpArt and Surrealism art movements but were interpreted with a less formal discipline. China Whites, Chrome and glossy lacquers emerged as important finishes during this period.

In 2007, Volkswagen created the ultimate Dream Barbie Pink Beetle in their Puebla, Mexico plant. The car was designed to help celebrate the Barbie Doll's 50th birthday that was rapidly approaching in 2009. The exterior was a light red with a pink iridescent luster pigment. The interior seating of the convertible was bubblegum pink and white quilted leather with the characteristic "B" monogrammed into the seats. The dashboard and cockpit were designed with pink and lipstick red surfaces, contrasted with black appointments. Only about a dozen of these cars were ever manufactured.

In 2008, Elon Musk launched his first Tesla roadster which was a luxury, high-performance green vehicle in more than 30 countries. The roadsters were offered in high-intensity shades of blue, orange and green. The second generation of these vehicles was much more conservative in shades of gray, red, white, black, brown, green and silver. The Tesla emerged as the new status green car, replacing the Toyota Prius.

Color Trends and Selection for Product Design.
DOI: http://dx.doi.org/10.1016/B978-0-323-39395-9.00012-8

Also in 2008, the Detroit Electric car was revived by the former Lotus CEO, Albert Lam. Detroit Electric started in 1907 and manufactured all-electric vehicles that were powered by an acid-lead battery that was rechargeable. Despite major competition from the more powerful, internal combustion-powered engine vehicles, Detroit Electric continued to manufacture electric cars until a few years after the stock market crash of 1929. Most of the car owners were women and medical doctors who needed a reliable car that started quickly without cranking by hand. The new Detroit Electric is expected to launch their SP-1 in 2016 and will sell for approximately $135,000 USD. The car will be manufactured in Europe.

Color Trends in 2010 for fashion and cosmetics contained the usual luxury shades of gold accents with periwinkle, russet, copper and bronze. There was also a unique shade of gray with a green tint, reminiscent of a heavy celadon. This was an interesting year in color trend because it was the first time in several years that fuchsia was paired with chartreuse. These are very dissonant colors and were attributed to the phantasmagoria trend that was sweeping the couture fashion world that was filled with anorexic models and over-styled fashions. These colors were pared with shades of emerald and navy that helped to tone down this pallet for the ready-to-wear market. Household appliances and hardgoods were following a trend of whimsical toy-like colors in highlighter colors of lemon, orange and cherry. These colors were anchored with stark blacks and white shades that would become popular in automotive in the following years.

In 2011, Lego launched a massive marketing campaign to target their lego bricks to girls with the introduction of the Lego Friends line. The line included more detailed female minifigures and the color scheme was coordinated in pink and purple shades with a suburban story line that includes a café, a riding stable, lemonade and ice cream stands and a convertible car. This series was one of Lego's most successful in their 65 year history Lego tried to market their bricks to girls in 1972, 1991, 1994 and 1997. The home maker series launched in 1972 emphasized rooms in the home, a doctor's office, a school classroom and a secretary's desk; however, Lego did not change their color scheme of the bricks and sold them in their traditional red, blue, white, black and yellow color scheme of the time. Although the set was on the market for 11 years, only 32 sets were released. The Belleville set, introduced in 1994, contained larger articulated dolls and pink and purple bricks, some embedded with glitter. In 2007, Mattel had announced three separate recalls of children's toys that were painted with lead paint in China. When the subcontractor Chinese factories were interviewed, they countered that they needed to use lead paint because it yielded brighter colors that were essential for children's toys and that the low margins offered by Mattel made it necessary to use cheaper lead-based paints to remain profitable.

The color trends of 2011 were heavily influenced by the 20th anniversary of the destruction of the Berlin wall, separating Western and Eastern Europe and the 10th anniversary of September 11, 2001 terrorist bombings in the United States. The European color trend influence took a long time to develop because the wall had been in place for about 40 years and had suppressed the economic development in Eastern Europe that had little to no influence on or from the rest of the world. The colors were young shades of industrial Bauhaus and earth tones but were also infused with a cautious optimism with some brighter colors. Industrial shipping container colors of blue, red and orange were brightened with coppers and silvers. A high tech pallet of graffiti colors of gloss black, sparkle silver was brightened by a fluorescent green and a tyrian purple. Both of these pallets were gender-neutrals. Another pallet of refined shades of money colors of refined pewter, aged gold and a gentian violet was reminiscent of the slumbering aristocracy that was re-awakening from the new economic opportunities flooding in from the west.

Lipstick colors grew brighter and bolder in 2012. This was also a time where there was an emphasis on ombre shading in nail enamels, fashions and hair color where there is a gradation in shade from light to dark. These ombre shades and techniques of application were directly linked to You Tube videos where consumers can learn new methods of application and try with little or no risk in the privacy of their own homes.

Color trends became more luxe in 2012 and were influenced by obscure regions such as Southeast Asia, Brazil and Australia. The Asian colors were more natural spice and silk colors that included sophisticated shades of heather and periwinkle. This is in sharp contrast to the traditional bright reds and yellows that generally dominate the regions. The on-coming World Cup Soccer tournament headed for Brazil in 2014 produced a slew of branding and advertisements that received global media attention about Brazil and resulted in a color pallet of vibrant oranges, greens and blues offset with linen shades of white and ecru that were quite important for sportswear and men's fashions. American luxe colors for 2012 were aligned with food trends of gourmet chocolates and coffees as well as potent superfoods such as pomegranate, salmon, beets and sweet potatoes. European luxe colors for 2012 were focused on the super-rich, oil-laden countries of the North Sea. Refinery colors such as bronze and steel were softened with wood tones and icy whites.

Teletronic colors of 2012 were less technical and more emotional and playful. Colors were bright but whitened shades of green, yellow, pink and blue and were important for driving cellphone colors in 2013 that previously mirrored automotive shades of black, white and silver. American ready-to-wear colors were heavily branded and influenced the presidential inauguration of Barak Obama, although it was not the president's fashions but rather those of Michelle, Sasha and Millea. Michelle's 2012 inaugural gown by Jason Wu was a stunning halter neck ruby chiffon dress that was in keeping with her refined sense of style evidenced by her 2008 inaugural white silk chiffon one-shoulder, floor-length gown that shimmered with Swarovski crystals. Sasha and Milea wore ready-to-wear overcoats in shades of blueberry and violet and the First Lady's purple leather gloves became emerging colors in fashion and subsequently housewares in 2013.

Although the "Kate Middleton Effect" officially started in 2010 with the announcement of her engagement to Prince William, her influence on style continues to current day. One of the more remarkable color trends that were attributed to the Duchess of Cambridge was when she wore tight coral-colored denim jeans to the Olympic Park in London for a hockey game in the spring of 2012. The Daily Mail in the United Kingdom stated that the sale of coral-colored denim jeans soared 88% in just a few hours of spotting the Duchess in the jeans. Her influence spread to other British celebrities who subsequently followed the trend. Her sister Pippa, Kate Alba and Frankie Sanford were all tracked in the following weeks wearing cobalt, yellow and bright green denim jeans as well. A similar effect is seen with her son George and his fashion influence is known as the, "Prince George Effect." The Royal Baby was ranked #49 on GQ's list of best dressed men in the United Kingdom for 2015. Some of the "Prince George Effect" was not only about the baby but also about Kate's pregnancy. As an example, Kate did not hide behind shapeless maternity dresses but rather favored more form-fitting tailored outfits that celebrated her pregnancy rather than hiding it. Also, the swaddling blanket that the baby was wrapped in on his first public appearance was made by aden + anais. The blanket was white cotton with little olive green primitive birds. It was part of their Jungle Jam series sold in a pack of 4 for approximately 60 USD. The Prince has also been photographed on numerous occasions wearing denim overalls or pants. Denim has been waning in boys wear for the last decade or so but seems to be returning at least for infants. He has also been spotted wearing navy blue Crocs that received a great deal of media attention.

Further Reading

https://en.wikipedia.org/wiki/Volkswagen_Beetle
https://en.wikipedia.org/wiki/Lego_Friends 2012
https://en.wikipedia.org/wiki/Car_colour_popularity 2012/2013
http://www.nytimes.com/2007/09/11/business/worldbusiness/11lead.html?pagewanted=1&_r=1 9LEAD PAINT
https://autouniversum.wordpress.com/2009/11/16/a-history-of-automobiles-and-color-before-the-age-of-chromophobia/
http://gas2.wpengine.com/wp-content/uploads/2014/05/tesla-colors.jpg
https://en.wikipedia.org/wiki/Tesla_Roadster
http://pinktentacle.com/2009/08/riba-robot-nurse-bear/
http://www.nytimes.com/2009/08/23/magazine/23Women-t.html?pagewanted=4&_r=0#
http://www.huffingtonpost.com/2013/08/04/prince-george-swaddle-aden--anais_n_3703740.html
http://news.bbc.co.uk/2/hi/uk_news/magazine/7817496.stm

13 2013–2014: The Slow Comeback

Pinks returned to the couture market but not by traditional means. As discussed previously, pink has been traditionally noted as a feminine color for the past 50 years but it was male athletes who brought the color back and populated it into women's fashions and hard-lines as they wore pink on the playing field to demonstrate their support for breast cancer awareness. Although pink crept into men's fashion, it was almost entirely removed by the end of 2013. Lolita fashions became firmly entrenched in Japan in a variety of subcultures. Sweet Lolita is a style similar to the Edwardian styles of the early 1900s. Unlike the early 1900s styles, Sweet Lolitas favor infantilized styles of young girls dressed in shades of light peach, pink and blues. American sportswear was dark and punctuated with accents of fluorescent green, orange and yellow. European colors were colder glacier or architectural green glass colors were combined with brushed steel and mercurized silvers, contrasted with stark whites and rubbery black effects. Clothing styles were more technical and electronic wearables became more mainstream in ready-to-wear fashions. A good example of this is the fit-bit and the jawbone which are rubberized wrist wear that is worn to monitor fitness levels. Fitbit was launched in 2008 as a clip on device in black and white and the Fitbit flex was launched in 2013 as a rubber wristlet and was offered in white, black and turquoise. In 2015, Fitbit offered the same technology as a designer series that can be inserted in metal or leather bracelet for double the price. Metal bracelets are offered in yellow or rose gold or silver, and leather bracelets are offered in white, tan, black or gray.

Middle-eastern and Latin colors were quite similar and were quite muted and modest in shades of bone, light coral and seafoam. Textures and finishes were matte and dry. Fashions were layered and modest in refined whites and beiges.

Asian and African colors were ostentatious shades of golds and coppers combined with indigenous bright colors of orange and blue. This pallet was muted by exotic wood, burgundy and red wine colors.

Automotive colors in 2013/2014 were still in the Silver, White and Black rut of the past decade. However, in 2013, Forbes reported distinct

Color Trends and Selection for Product Design.
DOI: http://dx.doi.org/10.1016/B978-0-323-39395-9.00013-X

gender preferences were uncovered where white was most favored by women, followed by brown and gold. Men were most likely to favor red followed by orange and black. Women tend to buy more family-oriented cars like minivans and SUV's whereas men favor sports cars and two-seaters in performance colors. There is a glimmer of hope in the interior of these vehicles where, although the surfaces are wildly monochromatic. Hints of colors are appearing in the over-stitching of seating surfaces as well as a variety of surface finishes on the instrument and door panels including brushed chrome, carbon fiber and sueded effects.

After a very public outcry of gender bias in 2013, Hasbro released their 50th anniversary Easy Bake Oven in a black, silver and blue color scheme to market the product to boys as well as girls. However, the 2014 version has been heavily marketed to girls with a purple, pink and turquoise color scheme.

A no-makeup trend started in 2014 where celebrities such as Gwyneth Paltrow and Giselle Bundchen participated in Instagram photo campaign with the hashtags "#No Makeup" and "#Makeup Free" to celebrate their natural beauty. This freedom from makeup could be seen as a backlash to the culture of perfectionism that is essential to the success of these celebrities. However, not everyone is confident enough to live life without the protection of makeup and white was a color that dominated eye shadows in 2014. For darker skin tones, this could translate into a pink or light brown shade that is still in keeping with the no-makeup trends. A very important lip color was a blue-shade pink that exhibits a soft purple with a pout of pink. This was also complimented with a very bright green shade of chartruse eyeshadow that would almost serve as a counter-trend to the no-makeup trend.

Further Reading

http://www.popsugar.com.au/beauty/Makeup-Free-Celebrities-36355679#photo-36355654 2014

https://www.makeupgeek.com/makeup-basics/3-spring-makeup-trends-and-how-to-wear-them/ 2014

14 2015–2016: Current and Future Color Trends

Once again, we see that society is experiencing chromatic fatigue. In 2014, we saw many dark shades of navy, black and olive punctuated with super bright blues, oranges, yellows and greens. In 2015/2016, we are looking at softer pinks, blues, greens and yellows combined with white and gold pearlescent, almost frosted colors. These are very young colors and very strong in North America. There is a retrospective influence of Expressionist Architecture from the 1920s that emphasizes natural phenomena such as crystals and rock formations, caves and translucent minerals that are illuminated from their interiors. European colors are quite different and are more focused on bold primaries of blue, red and yellow sharply contrasted with whites and blacks. Kitchen trends are also quite natural and will continue for the next 2–3 years. Colors such as moss green, avocado and a cleaner harvest gold will combine with dark wood tones and polished synthetic and rubberized surfaces. Variegated textures and colors will mimic root and exotic vegetables such as carrots, artichokes and eggplant. Asian and African colors are still quite luxe and metallic; however, there is less of an emphasis on silvers and golds for that region but are more focused on color-shifting effects, like peacock and ostrich, combined with umber shades of ocher, raw umber and a light pink iron oxides and contrasted with muted tones such as camel and cordovan. These muted tones are generally presented in smooth and polished leather finishes as well as highly lacquered black and russet. Automotive colors continue to stagnate in shades of white, gray, silver and black. There are some glimmers of color that continue to peak through in niche and luxury vehicles including bright greens and yellows and weather-stable dirtier orange shades and purples.

Future fashion and cosmetic colors and styles will be influenced, in part, by the Hijabers. Hijabers are Muslim women who wear traditional Hijab scarves. Hijab.com is a photoblog written by two Indonesian Muslim women who have started their own magazine for young Muslim Women called Laiqa where they discuss and display fashion

Color Trends and Selection for Product Design.
DOI: http://dx.doi.org/10.1016/B978-0-323-39395-9.00014-1

and how global fashion trends impact Muslim women. As Muslim values continue to modernize, many young Muslim women aspire to look at themselves through a combined lens of their education, their religion and social media and Western values. There is a great divide in the use of cosmetic products in this community where some Muslim women feel that cosmetics should be used sparingly and only for their husbands and others feel that they need to wear more flamboyant products in other social situations to gain the respect of their peers. This also points to the need for Halal approved colors and cosmetics so that they will not need to be removed for religious ceremonies and prayers. This is particularly important for permanent cosmetics that cannot be removed. This includes hair dyes. The current trend that beauty manufacturers use science to sell their products is largely unimportant to Muslim women who focus more on inner beauty and humility rather than glamour and technology.

The ideas around wearable technology continue to emerge and will employ color pallets that consist of high-tech metals, green and natural leathers as well as hints of more fashionable green-chic colors such as precious metals and gems, golds, platinum, diamonds and healing minerals in shades of pinks, greens and oranges.

Future automotive colors are predicted to remain uninspired in the coming years. This is due, in part to the extensive multiyear weather testing needed for exterior paint colors. Another factory is tightening of regulatory restrictions on the types of paints that will be permitted such as lowering the volatile organic components of the paint chemistries. The automotive industry is favoring the use of water-borne paint systems that are more environmentally friendly but are more difficult to apply and are unfriendly to certain types of pigments that are commonly used in solvent-borne paints. There is certainly a trend emerging to reduce the overly glossy effect of the Class A finishes that were mandatory in the last 20 years and replacing it with a matte finish. Automotive manufacturers have been highly resistant to this trend, claiming the car will look dirty and that the finish will be less durable. The reality is that this finish is quite expensive and that the paint chemistry has not caught up with the matte trend yet to create a durable and cost-effective paint. More of these matte effects are becoming evident in luxury vehicles and motorcycles and will eventually penetrate the mass-market vehicles. Another emerging trend is the car wrap. Although these full-body skins are mostly used for advertising on commercial vehicles, they are becoming more fashionable and artistic. Most of these skins are made from vinyl films. Although there are many photochromic and thermochromics exotic pigments currently on the market, it is unlikely they will be used in exterior automotive paints due to their lack of weatherability and their exorbitant price tags. There may be a use for some of these colors and effects in after-market vehicles and in accent areas in higher- end mass-production vehicles. Other automotive trends, however, will influence appropriate colors to meet those technologies. Greener technologies that create low- or no-fuel vehicles, smaller cars or cars that can fold or expand to meet increased space restrictions in urban environments and vehicles that can accommodate aging and increasingly disabled populations as obesity and mortality rates increase. At the end of 2015, Consumer Reports issued their list of most satisfying and least satisfying cars as rated by their owners. The top 3 most satisfying cars are the Tesla Model S, the Chevrolet Corvette and the Porsche Cayman. The top 3 least satisfying cars are the Kia Rio, the Nissan Sentra and the Jeep Compass.

There are seven current colors for the Tesla Model S: black, blue, red, gray and silver metallic, white pearl and flat white. The ten current Chevrolet Corvette colors are: black and white, two blues, two reds, two silver metallics, yellow and metallic orange (rust). There are 12 current colors for the Porsche Cayman: three shades of metallic gray, two shades of black and white (one metallic, one flat), brown metallic, lime gold metallic, red and yellow.

There are four current Kia Rio colors: metallic blue, yellow, white and metallic silver. There are four current Nissan Sentra colors: white, metallic silver, metallic gray and black. There are eight current Jeep Compass colors: black and white, two metallic grays, metallic silver, pearlized red, green and blue.

Further Reading

http://www.autobytel.com/jeep/compass/2015/color/

http://www.kbb.com/car-colors/2015-porsche-cayman-coupe-colors/

https://www.teslamotors.com/about

http://www.genengnews.com/gen-articles/animal-derived-agents-in-disposable-systems/1090/

http://www.rediff.com/business/slide-show/slide-show-1-auto-armadillo-t-a-car-that-folds-in-half-for-easy-parking/20130905.htm

http://www.hijab-scarf.com/2013/01/laiqa-magazine.htm

http://www.inspectorinsight.com/asia-trends/the-feeling-of-muslim-beauty-is-the-beauty-myth-universal-or-do-local-truths-prevail/#more-3226

http://www.9news.com/story/money/cars/2015/12/30/consumer-reports-list-owners-survey-satisfying/78076728/

SECTION II

The Processes of Color Trending

15 The Development of Color Trends

Traditionally, color trends are thought to cycle every 7 years. This is due, largely, to the extensive long-term exterior heat and humidity testing in Arizona and Florida for the automotive industry that requires several years to complete. Neutrals or classic colors such as blacks and whites tend endure over trend periods and are also statistically the best sellers in most industries. This is particularly true in more durable industries such as large appliances, exterior architecture and automotive. There is a certain value attached to each of these industries and the greater the resale opportunity, the more conservative or classic colors will dominate. Clearly, industries such as fashion or toys that are very immediate in their consumption and will have little opportunity for resale allow for color choices that can be more whimsical (fad-like) or reflective of current social and economic trends. Beauty products, including their packaging, tend to cycle more quickly with the fashion market so major colors change with the seasons or the annual fashion shows. This is also true, to a lesser extent to other types of personal care products and packaging. Female and children's products cycle more frequently and men's personal care tends to lag behind and has a much more limited color pallet that feels more automotive or sporty. That is not to say that these other industries do not pay attention to current social trends. As an example, prior to 2006, exterior automotive colors were quite broad and employed many effect pigments that were used in expensive tri-coat paints where a base coat of color was over-sprayed with a color-shifting or pearlescent pigment that was subsequently sealed with a glossy over-coat. After the fall of the stock market in 2008 and the subsequent great recession, automotive companies scaled back and a white, gray and black pallet dominated the market. Almost all of the pearlescent and metallescent color effects disappeared and were replaced with matte colors and limited amounts of aluminum flake to create a pallet of silvers and grays. The exception was the glass flake pigments coated with titanium. This was a multicolored sparkle pigment with pin-points of brilliance that was combined with white, gray and most notably black to create the still popular, "Tuexedo Black" in many Ford vehicles.

Color Trends and Selection for Product Design.
DOI: http://dx.doi.org/10.1016/B978-0-323-39395-9.00015-3

Different industries research trends differently. Even within the same industry, different researchers obtain their information differently. As an example, in the fashion industry, trends change every 6 months or so. Fashion week for ready-to-wear (RTW) designs is held every Autumn for the Spring/Summer fashions of the next year and in the Spring for the next year's Autumn/Winter colors. In order for designers to produce fashions for these shows, they will need to know the trends at least 6 months ahead of these shows so they can prepare their garments. The Haute Couture show only occurs in Paris once a year. This show only features women's fashion and is held in June after the Spring/Summer menswear show. The couture fashions are only shown at the fashion week in Paris due to Chambre Syndicale rules. The Chambre Syndicale de la Couture is a regulatory body founded by Charles Worth in the late 19th century that protected fashion houses from having their designs copied by lesser-skilled craftsmen. The current function of this association is to designate which design houses meet the standards of being designated as a Couture fashion house and participate in the Haute Couture show in Paris. The membership in this prestigious club has declined over the years. In the 1940s, there were over 100 qualifying Couture Houses. By 2005, there were less than 10. Even though the number of qualifying houses is very small, their impact on fashion trends and color trends is enormous. The list for 2012 includes: Adeline André, Chanel, Christian Dior, Christopher Josse, Franck Sorbier, Giambattista Valli, Givenchy, Gustavo Lins, Jean Paul Gaultier, Maurizio Galante and Stéphane Rolland. Foreign correspondents include: Elie Saab, Giorgio Armani, Valentino and Versace. There are very strict guidelines for membership including: custom designs for private clients, with at least one physical fitting, a workshop (*atelier*) in Paris with at least 15 full-time staff members, a minimum of 20 full-time technical people, in one workshop or more, two collections presented annually of a minimum of 50 original designs to the public of both day wear and evening wear. This couture level of fashion is frequently at the fore-front of color trends; however, this changes as industries rise and fall. As an example, the automotive industry was leading this front in color trends in the early 2000s prior to the Great Recession when the dominant car colors slid back into black, white and gray/silver. This was also true in the affluent 1950s when automotive colors flourished and pallets became bright and effusive as discussed previously. Fashion houses, like other industries, get their trend information from multiple sources. Some houses use confidential trend agencies whose only job is to scour the world for emerging trends in the art, technology and social arenas to distill new pallets. This can be quite expensive.

Other houses do their own research, creating teams to attend trade shows, art openings, scientific and social conferences to formulate their own pallets from their own findings. This method is sometimes criticized as being less objective and somewhat self-serving.

Generally, RTW designers use colors that are rolled out of the Couture Show in June in Paris. However, each designer will add their own preferences to their pallets and tailor it to their markets to use in their shows during the bi-annual fashion week shows around the world. We also see that mass-market clothing manufacturers, automotive manufacturers, cosmetic manufacturers, electronics designers and industrial designers use these global fashion RTW shows to create their pallets for their seasons, campaigns or year of designs. This is not a cut and dried process. Most companies that publish their own color trends use additional information from their own sources of inspiration which are used to round out these larger color trends. Additionally, most large companies that manufacture colorants, sell colorants to brand owners or downstream convertors that identify color as an important feature of their products (such as cosmetics or automotive) employs a color guru that collects and distills this information for their specific markets. Many of these people are well-known and respected in their industries and there are usually a handful of people in each manufacturing segment and serve in key roles in color organizations and sit on boards that select colors for specific industries.

Previously, color pallets were easily separated by geography. American fashion colors were quite distinct from European colors where European colors tended to be more conservative favoring classic neutral tones and blacks and whites where as American colors tended to take more risks and were much more effusive. Currently, we see that colors tend to be more globalized, being over-shadowed by brand preferences rather than geographical ones where we see a specific design house selling the same color schemes all over the world rather than catering to specific geographies.

Further Reading

http://www.modeaparis.com/en/news,4/
http://www.hongkiat.com/blog/50-trend-setting-design-studios-agencies/
http://www.dexigner.com/directory/cat/Packaging-Design/Companies.html
http://www.entrepreneur.com/article/175428
http://www.brandingstrategyinsider.com/branding-and-colors

16 Fads Versus Trends

The concept of a fad versus a trend may be difficult to separate in current times. It is generally thought that fads are a short-lived passion and trends are more of an enduring statement related to social, ecomomic and political occurrences. There are many examples of fads that have been endured over the years: pet rocks, jeggings, beanie babies and cabbage patch dolls, scrunchies, silly bands and the mullet haircut. In retrospect, these must-haves seem somewhat embarrassing due to their superfluous nature but in the moment, ignited a passion in the consumer culture, causing products to fly off the shelf and new ideas to be embraced by different cultures. These fads do not only relate to fashion and toys but also creep into other areas such as real-estates (time-shares, sky lights, sunken living rooms, mosh pits), interior home fashions (vertical blinds, shag carpeting, bean bag chairs, entertainment centers, track lighting) and hair styles (C). There are many fads related to hair color. In the 1950s, there was a burst of violet hair color that emerged and became popular among A-list celebrities in the 2010–12 timeframe. There are also some interesting hair fads for men, the scruffy Don Johnson look, the completely shaved head and in earlier times the toupee. In the 1970s, we saw an emergence of frosting and in the current decade the ombre or gradated hair colors that are very light on one end and very dark on the other. Currently, there is a distinct dividing line between older/middle-aged men versus younger men where we see that many men aged under 40 are not clean shaven but mustaches seem to fade in and out. There is also a lack of balding hair styles (such as the comb-over). Many men with hair loss tend to shave it all off or resort to high-tech methods to promote growth such as medications or hair grafts or weaves. In general, Western men are becoming more hairless on other parts of their bodies, opting for waxing, lasering or shaving similar to what women have practiced for decades. Automotive fads such as white-walled tires, pin stripes, synthetic wooden side panels and laundau vinyl tops. The emerging car wraps might fall into this fad as well if they do not evolve into more mainstream functional coverings. Kitchen fads such as banquettes, islands,

Color Trends and Selection for Product Design.
DOI: http://dx.doi.org/10.1016/B978-0-323-39395-9.00016-5

breakfast nooks and bars, tiled, granite and cement countertops will also continue to fade from popularity.

Further Reading

http://www.babble.com/products/kid-toys-zhu-zhu-pets-silly-bandz-gak/
http://www.examiner.com/article/celebrity-hairstyle-fads
http://brobible.com/guyism/article/awful-mens-hairstyles-haircuts-fads/
http://www.foxnews.com/leisure/2012/08/12/fads-from-180-10-to-remove-before-selling/
http://ezinearticles.com/?The-Latest-Kitchen-Designs---Weeding-Through-The-Passing-Fads&id=6019422
http://strongautomotive.com/worst-automotive-fads/

17 Macrotrends

A macrotrend is a large-scale, sustained shift in consumer interest or activities that can occur over several years or decades. These are usually viewed as "big picture" themes and are frequently used to make strategic business decisions and to justify the path that a company takes when they choose target markets, determine profitability or attractiveness, innovation strategies, mergers and acquisitions. Most companies allude to or directly refer to these macrotrends in their annual reports to their investors. Military and Government organizations also use these trends to chart or justify their activities in the past, present or future. Some examples of current global macrotrends are urbanization of population, aging populations, energy crisis and global warming, food shortages and terrorism. A typical method used to divine these macrotrends is a STEEPLED assessment. STEEPLED is an acronym for Social, Technological, Environmental, Economic, Political, Legal Ethical and Demographic analysis. These bigger trends can be distilled to further chart a strategies such as SWOT analysis (strengths, weaknesses, opportunities and threats) or Five Forces; a competitive strategy developed by Michael Porter. In 1970, Bruce Henderson developed the Boston Matrix which analyzes the business units or product portfolios within the structure of a company. Obviously, color trends are a smaller subset of these trends and can be, in part, derived from these broader Macrotrends. As an example, the Urbanization of Population macrotrend which points to the trend of growing cities was important in fashion color trends in 2010/2011 where clothing was predominantly styled in dark shades of black, gray, blue and red and punctuated with hints of bright primaries. City life tends to be quite dark and devoid of open spaces. Clothing, architecture and urban transportation tend to mimic these colors. These color trends were not only important for city dwellers but also became popular for suburban and rural consumers in months to follow. The use of highly chromatized primaries as accent colors continued to grow over the next following years and cannibalized the darker shades for a brief period.

Color Trends and Selection for Product Design.
DOI: http://dx.doi.org/10.1016/B978-0-323-39395-9.00017-7

Macrotrend: Food Shortages

Food shortages or a poor distribution of food throughout the world is another macrotrend where there is a link to color trends. In 2012, there was a surge in global exotic food colors that influenced the pallets of home goods, furnishings and small hard goods. Many of these were spice colors like saffron, cumin and dried herbs and edible flowers such as sage and lavender. This color pallet resonated with global citizens of the world who enjoyed eating exotic flavors from markets and ethnic cafes. These venues taught consumers how to appreciate smaller amounts of food by making them more flavorful and colorful. It also expanded the western diet beyond meat and potatoes and ushered in a whole host of alternative protein sources including insects and plant-based proteins as well as nondairy alternatives including soy and nut milks to replace traditional bovine milk, cheese and yogurts.

Macrotrend: Aging Population

The aging population ushered in a slew of high-tech health care products to accommodate the challenges associated with the elderly including hearing impaired devices that looked like earrings or provided supertitles inside of eyeglasses streaming across the lenses. Google developed contact lenses that could detect insulin levels of the wearers who suffered from diabetes. These colors tended to be quite metallic and powerful, focusing on their elevated abilities through these technologies rather than their disabilities from aging.

Macrotrend: Global Violence and Terrorism

Global terrorism and the collateral damage associated with these events also promoted a similar high-tech pallet. The development of the Blade prosthetics that allowed the user to run and jump faster than their noninjured counterparts. A company in Korea developed robotic wheel chairs that placed the double amputee into a waist-sized socket that enabled the injured to move around at an elevated position, at eyelevel with most able-bodied consumers while turning and propelling themselves at elevated speeds and still remaining hands-free for holding a cellphone or briefcase. The colors were primarily masculine metallics of silver and stainless accented with blacks anodized metals and glass greens.

Macrotrend: Eco Crisis

Global warming and the energy crisis created new technological solutions and the colors that emerged from them were quite significant as well. Elon Musk, visionary inventor associated with the electric car, the Tesla, has proposed a solar high-speed train that would travel from New York to Los Angeles in just 2 hours. Another company commercialized a bracelet that would detect UV rays and alert the wearer via an app that can be downloaded to their smart phones to indicate that they needed to cover up their exposed skin or apply sun block. Obviously there are many shades of green and earth tones, natural leather and wood colors in this pallet that were important in fashion in 2015 and 2016.

Further Reading

http://yourbusiness.azcentral.com/macro-trend-8289.html
http://www.slideshare.net/lconner/steeple
https://en.wikipedia.org/wiki/PEST_analysis
http://www.quickmba.com/strategy/porter.shtml
http://www.marketingteacher.com/boston-matrix/
http://www.endolite.com/products/elite-blade
https://www.teslamotors.com/blog/secret-tesla-motors-master-plan-just-between-you-and-me

18 Aging and Healthcare

As indicated earlier in the Macrotrend section, the aging population demographic plays a major role in shaping new business opportunities, products and services. This is especially true in the medical, mobility and pharmaceutical industries. Not only are many drugs focused on prolonging life, many of the delivery devices and packaging are geared for the senior population.

Packaging is an example where modifications in child-proof packaging has also created senior-proof packaging where older consumers with arthritic fingers and weakened joints find it difficult to squeeze or manipulate this sophisticated packaging. This has also generated a new segment of packaging known as Senior-friendly packaging. Also the use of brighter colors, makes the packaging easier to see or recognize for the visually compromised. Japan is a hotbed for robotics development and currently one in four Japanese is currently 65 or over and is expected to reach 40% by 2050. Toyota is pursuing this mega trend and expects to create a healthcare robotics market in Japan that will be a core competence by 2020. Tokai Rubber Company and RIKEN developed a friendly looking robotic nurse that has the face of a bear and is covered with a soft-touch polymer that is more comforting when in contact with human skin. RIBA (Robot for Interactive Body Assistance) is designed to lift patients in and out of bed and perform some other basic nursing skills. The bear can recognize faces and respond to verbal directions. Although the exterior is mostly white, it has a light blue neckerchief and a skirt that flares out from the bottom of the robot that gives the appearance of both cleanliness and calm.

Brighter colors are also used to make delivery devices more appealing and less clinical looking and are proven to enhance consumer compliance in correct usage. This is true for the younger population of children as well. An example would be the use of brightly colored inhalers for young asthmatics who view these items as more appealing or fashionable because they appear more toy-like than medical devices.

Color Trends and Selection for Product Design.
DOI: http://dx.doi.org/10.1016/B978-0-323-39395-9.00018-9

Further Reading

http://pinktentacle.com/2009/08/riba-robot-nurse-bear

http://www.cpsc.gov/en/Regulations-Laws--Standards/Statutes/Poison-Prevention-Packaging-Act/Child-Resistant-and-Senior-Friendly-Packages-packaging-guide/

http://www.toyota-global.com/innovation/partner_robot/

19 Gender

Gender lines continue to be drawn more strenuously and at earlier ages in our society. As mentioned previously, most gender division started in the children's apparel market in the 1920s when Time Magazine stated that pink was more suitable for boys because it was considered to be a stronger color and blue was better suited to girls as it was viewed to be a more delicate color. This flipped in the 1940s when consumers began to favor blue for boys and pink for girls. This has sometimes been attributed to Hitler's use of a pink triangle as an indicator for homosexuals. This was almost exclusively attributed to male homosexuals because female homosexuals were mostly invisible to society at that time so the pink triangle was associated with femininity. Prenatal gender determination, through the use of ultrasound, has allowed parents to prepare for the new baby girl or boy that will enter their lives in the next few months. Generally, this is a positive move, so that families can prepare the nursery, diapers, clothing and toys ahead of the birth. The darker side is that families can selectively abort fetuses that are not the desired gender. This is especially true in China where sex-selection abortions have been banned but female adoptions are still quite common. This has created a shortage of females in Asia and especially China and a new generation of Flower Boys has arisen. Flower Boys is a term ascribed to Asian boys that are raised without siblings that are generally over-indulged. As a result, men's fashions and furnishings have become more flamboyant and provocative, in part, to attract the minority of eligible females in their society. Historically, men's fashions have been muted or dark in composition whereas women's fashions tended to be brighter, more flirtatious and provocative. There is evidence to suggest that men's fashion was intended to highlight their strength and power without calling obvious attention to their clothing whereas for women, brightly colored, elaborately decorated costuming has always been and remains in fashion not only to call attention to their female form but also a signal to their presence.

In general, men prefer brighter colors and their associated shades with black and women prefer pastel or softened colors tinted with white. Blue is a favorite for both men and women. Men's most favorite colors are

Color Trends and Selection for Product Design.
DOI: http://dx.doi.org/10.1016/B978-0-323-39395-9.00019-0

blue, green and black. Women's most favorite colors are blue, purple and red. Men's least favorite colors are brown, orange and purple. Women's least favorite colors were orange, brown and gray. Yellow tends to be the most visually irritating color for both genders. Men also tend to prefer simple color names (roygbiv) and women tend to prefer more nuanced names (periwinkle, magenta, slate, etc.). This might have something to do with the fact that women and men are predisposed to see color differently. In 2012, the Rhode Island Rams had an unfortunate color scheme in their football uniforms and helmets in shades of powder blue and white. Their school colors are navy blue, light blue and white but for some reason, they decided to minimize the navy and the uniform and helmet were primarily light blue and white. The helmet had a white swirl on each sides similar to many other teams using the same horn imagery. The difference was in the color scheme. The light blue helmet with the white swirls seemed quite feminine and contrary to the masculine prowess that is normally displayed on the football field. Clearly the combination of light blue and white is quite beautiful but some of its detractors felt that it weakened the image of the team.

In a study published by Abrams of Brooklyn College in 2012 regarding differences in color detection based on gender, Abrams found that women tend to detect shorter wavelength colors such as blues and purples better and perceive them as warmer whereas men tend to detect longer wavelength colors such as reds and oranges and perceive them as more vibrant. Women are also better at distinguishing the differences between nuanced shade colors whereas men are better at detecting absolute colors. Men are also better at detecting fast-moving objects at a distance and seeing the details of those objects. The theory is that men have more neurons in visual cortex in utero due to the fact that section of the brain is flooded with testosterone during fetal development. This plays into many stereotypes and over generalizations of men being better drivers and women being better at gathering (berry detection in the bush) rather than hunting. Color is acuity is a relative comparison and there are many men who are good at detecting nuanced colors and highly skilled women drivers so these are generalizations but not absolutes in gender differences.

There was also a shift in kitchen schemes, and appliances in the late 1980s and early 1990s when men picked up more domestic chores and the balance of work/family life shifted to a more even gender distribution. Prior to this, kitchen colors and domestic appliances were more organic greens, browns and golds but then changed to more masculine/neutral white black and silver.

There has been great resistance to the over-feminization of products. The automotive industry has tried on numerous occasions to create cars that are more feminine in appearance and have been largely unsuccessful. A good example is the Volkswagen Beetle that has been in production since 1938 when Adolf Hitler saw Josef Ganz's May Bug at an automotive show in 1933 and encouraged the production of the first, "Car of the People." In 1998, the Beetle was redesigned with a more rounded exterior and a retro flower vase that was available in the earlier 1950s models. This "New Design" was quickly panned by the press as a "Chick Car" and was given a muscular, less round design in 2012. The numbers support this claim since 60% of the buyers in 2010 were female. Another, more specifically female-designed vehicle was the 1955 Dodge the La Femme. The color scheme was "Sapphire White" and "Heather Rose" and the La Femme model nameplate was created in gold script by the front fender. Although Dodge provided a tremendous amount of marketing and publicity about these vehicles, they failed to sell which has been attributed, in part to the Dodge sales force that mostly male and more comfortable selling automobiles to other men. Recently, the Fiat 500 was dubbed as too effeminate and several commercials have been launched to point out the more masculine, beefier design of its exterior, including a vignette where an older man mistakenly drops his male enhancement pill into the gas tank of the Fiat that subsequently swells its fenders and speeds off into the sunset.

Further Reading

https://en.wikipedia.org/wiki/Prenatal_sex_discernment

http://www.nytimes.com/2009/08/23/magazine/23Women-t.html?pagewanted=4&_r = 1#

http://time.com/105292/gender-wage-gap/

http://www.nytimes.com/2014/02/23/opinion/sunday/on-campaign-trail-missteps-on-gender.html

http://www.nytimes.com/2013/02/17/opinion/sunday/why-gender-equality-stalled.html?pagewanted = all

http://articles.latimes.com/2013/dec/26/local/la-me-one-way-gender-revolution-20131227

https://en.wikipedia.org/wiki/Nazi_concentration_camp_badge

http://www.colormatters.com/color-and-vision/color-and-vision-matters

http://news.nationalgeographic.com/news/2012/09/120907-men-women-see-differently-science-health-vision-sex/

http://www.brooklyn.cuny.edu/web/academics/faculty/faculty_profile.jsp?faculty=482
http://www.post-gazette.com/health/2006/09/13/Some-women-may-see-100-million-colors-thanks-to-their-genes/stories/200609130255
http://www.forbes.com/power-women/list/2/
http://www.smithsonianmag.com/arts-culture/when-did-girls-start-wearing-pink-1370097/?no-ist=
https://en.wikipedia.org/wiki/Volkswagen_Beetle
https://blog.kissmetrics.com/gender-and-color/
http://www.rmki.kfki.hu/~lukacs/TETRACH.htm
http://www.post-gazette.com/health/2006/09/13/Some-women-may-see-100-million-colors-thanks-to-their-genes/stories/200609130255
https://en.wikipedia.org/wiki/Rhode_Island_Rams_football
http://bleacherreport.com/articles/1263439-college-footballs-15-worst-uniform-color-combinations/page/5
http://coschedule.com/blog/color-psychology-marketing/
http://www.colormatters.com/color-and-vision/color-and-vision-matters
http://www.colormatters.com/color-symbolism/gender-differences

20 Geography and Culture

Although, geography and culture are inextricably linked, it is important to look at these components separately. In northern, cooler climates, where winters are long and light is low, warmer colors are frequently favored. In desert areas where the light level is high neutral colors tend to disappear so tropical bright colors such as peacock green, purples, oranges, pinks and limes are favored for architecture and fashions. When selecting colors for a new product or design project, understanding how a color or color family is interpreted in a specific geography and by a culture is very important to the success of that design. For the sake of clarity, we will limit the discussion to the 20th and 21st centuries for context. However, it is important to note that these sentiments are changing rapidly due to the electronic globalization of cultures. As the Internet continues to illuminate and homogenize cultures, these differences are becoming more subtle but will continue to influence color preferences especially in older consumers and those who have less access to social media and cues.

In ancient Chinese the character for color is 色 which literally means the color of the face or could be translated as emotion. This might make sense for most consumers but for color chemists and engineers, color is frequently reduced to a series of codes and analytical measurements that are entirely divorced from sentiment. The most successful color manufacturers and designers try to incorporate both the rational and emotional components in their business decisions but frequently the rational side prevails in business decisions. In traditional Chinese culture, there are five basic colors, red, white, black, quing (green) and yellow. This theory, known as Fu Xing or the Five Elements, is a comprehensive philosophy that suggests the interactions of basic life elements. These five elements are Fire, Water, Earth, Metal and Wood. There is a level of granularity between these elements that is associated with color. The Fire element color is obviously red and to a lesser extent purple its direction is South. Red is traditionally the color of joy and prosperity and is prohibited at funerals. It also considered to be a masculine color. The Water color is black and to a lesser extent gray its direction is North. Black is also considered to be the colors of the

Color Trends and Selection for Product Design.
DOI: http://dx.doi.org/10.1016/B978-0-323-39395-9.00020-7

heavens and is also viewed as a neutral color. It is also considered to be a feminine color. The Earth element colors are yellow/orange and its direction is center. Yellow is the most prestigious color and is in the center of everything. It is also the color of good luck. It also considered to be a masculine color. Interestingly, yellow is also the color of mourning for Chinese Buddhists. The Metal element colors are white and silver and its direction is West. White is the color of purity and mourning and is frequently evident in funereal events. It is also considered to be a feminine color. The Wood element colors are green and brown and its direction is East. Green is viewed as a harmonious and prosperous color. It also considered to be a masculine and color of fertility. Green is sometimes associated with infidelity so care must be exercised when designing with green for the Chinese culture. Additionally, it might be noted that blue is absent from this philosophy. Blue was traditionally categorized with green under the Wood element but was named as a separate color in later times. Blue is considered to be a color of growth and health. Blue and green together can symbolize balance and stability. Blue gemstones were used to create talismans worn by superstitious pagans who believed they would ward off evil spirits and bad luck. Blue light is sometimes associated with inappropriate sexual activity or lewdness.

Conversely, in the United States and in other Western cultures, Black is associated with death, depression, power and sophistication. Reds, hot pinks and purples are colors of romance; however, in Northern Europe, pastel colors are viewed as the most romantic. Colors are also used to indicate political movements and unrest. Most of these are nonviolent and are related to the fall/decline of the Soviet Union in the early 2000s but there are examples of other color revolutions in other parts of the world as well. To symbolize their nonviolent nature, many of these revolutions used a flower color to organize the demonstrators who dressed in the same color shade.

Red Shades

Red shades are frequently viewed as the most provocative of all color families. Since it is also the color of blood, it is frequently associated with love, romance and courage. Since red is the most visible color in the daylight, and is universally indicated as a color of warning or danger it was chosen as the universal color for stopping traffic by the *Vienna Convention of Road Signs and Signals* in 1968. There are also many misnomers associated with red. As an example, bullfighters in

Spain wave a red flag at bulls to incite aggression but the reality is that the bulls are actually color blind and do not see the red shade at all and are responding to the movement of the flag rather than the inflammatory color. Another common misconception is about Communism and the Red Square in Moscow. The genesis of red in the Red Square comes from the Slavic word, *krasnaya*, whose literal interpretation meant beautiful rather than the literal color red that is typically attributed to war and bloodshed of the Russian Revolution in 1917. In 2003, Georgia staged their Rose Revolution to overthrow the reign of President Eduard Shevardnadze. Red is also considered the color for Sunday (Sun planet) in the Thai calendar whereas pink is indicated for Tuesday (Mars). Red and yellow are sometimes associated with dryness. Red is also the symbolic color of the Goddess Lakshmi. Lakshmi literally means "goal" in Sanskrit and she is depicted in red robes to symbolize success in spiritual and financial success. Her robes are trimmed in gold to indicate the importance of prosperity but it is still secondary to spiritual success. In 2004, the Tulip (pink) Revolution began in Kyrgyzstan to protest the corrupt practices of President Askar Akayev. The Lotus Revolution (pink) occurred in Egypt in 2011, when President Mubarak was forced to out of power as part of the Arab Spring. Nazis also used a red triangle to indicate communist or socialist prisoners. They also used pink triangles to indicate male homosexuals and purple triangles to indicate prisoners who were identified as Jehovah Witnesses. The Nazis chose to use the triangular shape because it is commonly used in German road signs to indicate caution.

Orange Shades

Orange is not a neutral color and usually evokes strong emotions of love or hate in many people. It is a universal, nongender children's color. In Hinduism, orange or saffron is a holy color whereas in the United States it is frequently used as a color for prison inmate uniforms. Buddhist monks where saffron robes since it is the color associated with humility. Orange is sometimes viewed as warm-climate or fruity color and also is associated with autumnal shades of pumpkins and changing leaves. In France, the middle caution light is orange. Orange is a high visibility color and orange fabric was used to make life jackets for pilots in World War II so they were visible while floating in water after a crash. It is still used for spacesuits, vests and jackets for highway crews and construction workers to increase their visibility and safety.

Aircraft black boxes are also orange to enhance their visibility after a crash. Although the Nazis did not use orange in their inmate marking system, they did use a brown triangle to indicate Romas or Gypsies. The Romas are people who have originated in northern India and settled in eastern Europe, Turkey and Spain. They are sometimes erroneously defined as Romans or Romanians. Orange is the color indicated for Thursday in Thailand and public telephone booths are orange in Malaysia. In 2004, the Orange Revolution occurred in the Ukraine when Viktor Yushchenko's supporters took to the streets in large-scale protest to corrupt elections practices in their country. The Saffron Revolution took place in Burma (Myanmar) in 2007 and was led by students and monks to protest the removal of subsidies on fuels that increase by over 500% in 1 week.

Yellow Shades

Yellow has many different meanings all over the world. As mentioned previously, in China, yellow is a color of nobility and joy. In Western cultures, it is one of the least favorite colors representing ideas such as exclusion, cowardice and betrayal. Yellow is the symbolic color of the God Vishnu. Lord Vishnu is the Buddhist deity that protects and sustains the universe. He has four arms: two for his earthly duties and two indicating his spiritual presence. He is dressed in yellow as a symbol of his earthly presence. The yellow star, marked the Jews in the Nazi concentrations. The double triangle shape was used to replicate the Star of David as well as indicate double caution in their badge coding system. In Islam, yellow represents wisdom. In medieval Europe, yellow represented reason. Globally, it is considered to be a color of visibility and caution. Yellow is the color for Monday (moon) on the Thai calendar. In 2005, Moscow's House of Cinema hosted a film symposium entitled, "The Yellow House of Cinema" that examined mental illness and its place in the Russian film industry. A "Yellow House" is a common reference for an insane asylum in Russian culture. Yellow shades of hair give the wearers advantages. The "Lemon Law" was enacted in 1975 in the United States that provides compensation for consumers who purchased poor quality, defective automobiles, trucks and motorcycles. The Lemon Law has been expanded to include housewares, electronics and heavy machinery. Blonde males and females get more hits on dating sites. Blonde female fundraisers obtain more donations than their brunette counterparts.

Green Shades

Green is frequently associated with nature and health. The roots of the name are found in grass and growth. It is indicated as the color for Wednesday (Mercury planet) in the Thai calendar. In Western cultures, there is an association with sickness and envy: toxicity and poison. A green cross indicates a pharmacy in Europe. In Thai, green is an adjective for a foul odor. The Japanese have two names for green. The first is *midoru* which is more closely associated with fruits and plants. The second is *aoi* which is a color attributed to the blue family. Green traffic lights in Japan are designated as *aoi* but are still a similar shade of green used in other countries. The Green Movement, or the Ecology Movement is a global emphasis on creating societies that are more earth friendly. This Movement officially started in the West in the early 20th century after the rise of the Industrial Revolution which impaired or destroyed many of the natural resources in those parts of the world. The Green Revolution started in Iran in 2009 to protest corrupt election practices. Green was chosen because this was the color of campaign signs that were used by Mir-Hossein Mousavi Khameneh who lost his first election due to corruption. Green is also associated with moistness. Green wedding dresses and automobiles are considered unlucky in some Western cultures. The Cedar Revolution in Lebanon is sometimes associated with the green cedar trees of Lebanon and began in 2005 to initiate the withdrawal of foreign troops (Syrian) from Lebanon. Green is also a color of money and wealth and can be related to the US dollar that was pigmented green to deter counterfeiting by photography. A green triangle was used by the Nazis to tag prisoners who were former gang members that were selected to be Kapos which were a select group of prisoners of elite status that worked with the SS forces inside the camps to enforce the rules and carry out many of the brutal tactics that the camps inflicted on the inmates that some of the German soldiers were unwilling to enact.

Blue Shades

Blue is an important color for fashion and art. Blue is indicated as the color for Friday (Venus) in the Thai calendar. Politically, it is the color for the conservative parties in the United Kingdom and Canada and the democratic (centrist and more liberal) in the United States,

Denmark, Sweden and the Netherlands. Kuwait staged a Blue Revolution in 2005 to achieve women's right to vote. Belarus staged the Blue Jeans or Denim Revolution in 2009 since Blue Jeans were regarded as a symbol for freedom. This revolution was a public demonstration of the Society's disenchantment with their voting practices and were characterized by people waving strips of blue denim as ribbons. Blue is also a color of purity in many religions. The emotional content of blue is generally associated with sadness, and depression and musical genres such as Blues or Blue Grass. Nazis used a blue triangle to indicate prisoners who were foreign laborers against their will. Blue is also associated with water and wetness in Western cultures. Blue is the symbolic color of the Indian God Krishna, the Hindu God of love (Krishna is derived from the literal dark blue), although artistic sculptures and paintings frequently depict him as having a light blue skin or wearing bright blue clothing.

Purple Shades

Purple is a combination of red and blue but depending on the balance, can quickly shift to gray. The genesis of purple is derived from the Latin *Purpura* as the natural color that is extracted from the mucosa of the Murex snail. This was the origins of Tyrian purple that is associated with intellect and royalty in many cultures. It was also a color that was reserved for the highest military and political positions and was frequently worn after successful conquests. This exclusivity meant that it was also banned in certain cultures as a color that was prohibited for use by the masses and a violation of this ban could lead to death by execution in Italy. Thus, children born to emperors were noted as being, "born into the purple" because they were permitted to wear purple as part of their birthright. Part of the reason for this exclusivity was related to the prohibitive cost to produce the natural purple. As mentioned previously, the dye is extracted from tiny snails. About 12,000 crushed snails yield a little over 1 gof pure dye. Once the snails are crushed, the hypobranchial gland is removed and the extract is exposed to sunlight until the color shifts from colorless to white, yellow to green to finally arriving at an enduring purple. This time-consuming process generally yielded about enough dye to color one garment in this Tyrian Purple shade. Purple shades are generally not popular in China and Japan. The Japanese sometimes consider purple to be an arrogant color although it is a color worn by their Buddhist monks. The Chinese tends to associate shades of purple with grief and weakness. In contrast, the US government awards the Purple Heart, a military medal, for bravery. In Thailand, purple is the color of Saturday (Saturn). Purple is also associated with death in Western cultures. Black and purple buntings are frequently displayed at funerals of police officers or fire fighters. The Christian Lenten season is also associated with shades of purple. The color tends to lighten to a more optimistic shade of violet for Easter, symbolizing the resurrection of Christ. In India, purple is regarded as a meditative color whereas in Latin America, many men consider the color effeminate and refer to homosexual men as "Lilos" (lilac). The Grape Revolution occurred in Moldovia in 2009 to protest unfair election practices. Grape was chosen because of the abundance of vineyards in Moldovia. Purple is also a common color used in luxury goods and packaging, especially chocolate. Sometimes purple is associated with gang colors and drug lords. Purple prose is considered flowery or overly embellished to the point that it overshadows the thoughts or ideas conveyed in the writing. The Velvet Revolution started in 1989 to demonstrate against communism and subsequently led to the collapse of the communist regime in Czechoslovakia. In 2005, the Purple Revolution started in Iran to protest corrupt voting practices. The color purple was chosen since, in Iran, voter's fingers are marked with purple ink after they vote to prevent them from voting more than once.

Black and White

Many color chemists believe that black and white are not colors but are merely the presence or absence of all colors. White can be split into all colors when passed through a prism. The Hindu's have seven different words for white, each one referring to a different shade: pearl white, silver white, milk white, moon white, sunlight white, starlight white and sandalwood white. Sandalwood white is prepared by grinding the wood on a granite slab and slowly adding water to create a thick paste that is used for religious ceremonies where it is applied to the skin to bring followers closer to the divine. White is frequently associated with purity, cleanliness and virtue. However, not all associations with white are positive. The racist group of the United States, the Ku Klux Klan wore white robes to symbolize their belief in white racial supremacy. In many Asian cultures, white represents death and sometimes reincarnation. There were two Jasmine (White) Revolutions that occurred in the 21st century. In 2010, Tunisa started their Jasmine Revolution in response to rid the country of President Ben Ali at the beginning of the Arab Spring. In 2011, the Chinese organized their own Jasmine Revolution to protest the Internet censorship that occurs in China. The White Rose was a resistance group that formed in Nazi Germany to protest violence. Queen Victoria's white lace wedding dress changed the tradition of brides wearing dresses of any color to completely white in 1840. Hindu Goddess Saraswati is the Goddess of learning and wisdom and is depicted wearing white robes and sitting on a white lotus because white is the color that is representative of learning and purity.

By some definitions, black is the absence of all color. However in printing, the CMYK designation, cyan + magenta + yellow = key or black. Truly the darkest color, black is traditionally associated with darkness, evil, death, mourning and depression in Western cultures. Black is also viewed as the most elegant color followed by metallic colors of gold and silver as well as white and white pearl. The Nazis used a black triangle to symbolize antisocial people including the mentally ill or disabled, prostitutes and pacifists, as well as lesbians. In 2000, Yugoslavia staged their Black Revolution (Bulldozer Revolution) to overthrow the dictator Slobodan Milosevic. The revolution was led by many of the laborers, many of them coal miners and when a bulldozer operator drove his equipment through a police barricade during the protest, the revolution earned its appropriate name.

Special Effect Colors

In addition to colors, it should also be noted that special effect shades and metallics also vary in their cultural perception throughout the world. Typically special effect shades or pigments lend luster, pearlescence, metallic reflectivity or color shifting capabilities to a finished product. Generally, effect pigments can be categorized into pearlescents, iridescents, metallescents and metallics. Effect pigment shape, size and morphology will influence the overall appearance of the finished product. Pigment size is usually characterized in microns. A micron is one millionth of a meter and an average particle size range for effect pigments is 8–48 microns. Smaller effect pigments have a tighter range in the 2–10 micron size or 4–32 microns. Some pigments have a broad range from 5 to 300 microns and there are larger pigments that are manufactured upwards from 1000 to 3000 microns. Smaller particle sizes create a smooth continuous effect, medium particle sizes create more of a distinct pigment, matte effect and larger particle sizes yield more pop and sparkle. Pearlescent pigments are usually associated with luxury and value-added products and white pearls are the most commonly used but shades that are more yellow (gold pearls) or more blue (silver pearls) are also popular for luxury products. Iridescent products are similar to pearlescent pigments due to their ability to create luster but they also change color on viewing angle. Color shades range from violet, blue, orange/red to green. A subclassification of iridescent pigments are OVPs (optically variable pigments). These pigments create many different color shifts on different viewing angles. These iridescent pigments are frequently used for high-tech effects. Metallescent pigments are those that appear to have a softened metallic effect and yield metallic shades of copper, bronze, russet, brass and silver. Metallic pigments are simply metals that are pulverized into small particle sizes that can be incorporated into paints, plastics and cosmetics. Their effect is brighter and more reflective than metallescent pigments and are sometimes seen as more of an industrial effect rather than more of an esthetic effect. It is important to note that designers and stylists use and judge these products based on their own views of their appearance. As an example, large particle sized effect pigments are frequently dismissed as glittery or cheap. However, recreational equipment and goods such as racing boats and fishing lures use these large particle metallic products to denote high energy products that have pop and sparkle or attract the attention of fish. Glittery and large particle sized effect pigments are frequently used in toys and disposable housewares and tend to cheapen the appearance of the product. Iridescent effect pigments are sometimes associated with cheap and flashy effects in semidisposable consumer goods but can also denote value-added quality and fashion forward design sensibility in shoes, handbags and consumer electronics. White, gold and silver pearlescent pigments are widely accepted as classic, richer colors but when used to boost the status of inexpensive or disposable items they may also appear as gaudy or fake. Care should be exercised to use these products in moderate concentrations or perhaps in a trim or accent piece of the finished design rather than applying it to the entire piece.

Further Reading

https://en.wikipedia.org/wiki/Fashion_week
https://en.wikipedia.org/wiki/Color_in_Chinese_culture
https://en.wikipedia.org/wiki/Color_Marketing_Group

http://www.colorcom.com/research/why-color-matters
https://en.wikipedia.org/wiki/Wu_Xing
https://en.wikipedia.org/wiki/Yellow
https://en.wikipedia.org/wiki/Environmental_movement
https://en.wikipedia.org/wiki/Green
http://www.colormatters.com/the-meanings-of-colors/orange
https://en.wikipedia.org/wiki/Red
https://en.wikipedia.org/wiki/Red_Square
https://en.wikipedia.org/wiki/Romani_people
https://en.wikipedia.org/wiki/Russian_Revolution
https://en.wikipedia.org/wiki/Thai_solar_calendar
https://en.wikipedia.org/wiki/White#White_in_other_cultures
https://en.wikipedia.org/wiki/Colour_revolution
http://www.koausa.org/Gods/God10.html
http://www.koausa.org/Gods/God3.html
https://en.wikipedia.org/wiki/Krishna
http://nvdatabase.swarthmore.edu/content/kuwaiti-women-struggle-suffrage-blue-revolution-2002-2005
https://en.wikipedia.org/wiki/Mir-Hossein_Mousavi
https://en.wikipedia.org/wiki/Public_copyright_license
https://en.wikipedia.org/wiki/Purple_prose
https://en.wikipedia.org/wiki/Rose_Revolution
http://www.rusfilm.pitt.edu/2005/
http://conradcommunications.com/work/writing/the-blonde-effect/
https://en.wikipedia.org/wiki/Tyrian_purple
http://nvdatabase.swarthmore.edu/content/ukrainians-overthrow-dictatorship-orange-revolution-2004
http://www.answers.com/Q/What_is_Krishna_god_of
http://www.inspectorinsight.com/semiotics/the-meaning-of-purple-the-colour-of-nobility/

21 Economic Status

Generally speaking, the more affluent members of society tend to favor colors that are muted and neutrals. This idea of understated luxury is intended to hide their affluence and not attract attention. Of course, in cultures that have recently ascended the financial ladder, colors can be overly bright or even garish and are intended to have the opposite effect of drawing attention to their newly obtained wealth and perceived status. The idea of attracting attention and minimizing is especially true in automotive colors. Sports and high-performance vehicles tend to be painted in high-visibility colors of red, orange and lime green. Most of these vehicles are purchased by men. Family cars such as minivans, SUV's and station wagons tend to be in warmer or more neutral shades of brown, gray and dark green. Super luxury, nonperformance sedans that emphasize privacy and comfort also tend to be painted in rich shades of burgundy, silver, sophisticated navy blue, glossy black, sporting green and cream. Universally, orange, yellow, brown and red (to a lesser extent) are considered to be cheap or inexpensive colors.

Further Reading

Bourdieu, P., 1984. Distinction: A Social Critique of the Judgment of Taste. translated by Richard Nice. Harvard University Press, Cambridge.

http://www.encyclopedia.com/article-1G2-3035200131/clothing-and-social-status.html.

Color Trends and Selection for Product Design.
DOI: http://dx.doi.org/10.1016/B978-0-323-39395-9.00021-9

22 Global Color Influences

Different regions in the world are known for traditional regional pallets that endure beyond time-limited trends. These pallets fade in and out of global trends depending on political events, economic prosperity, religious influences and social unrest.

Asia and Africa

Earth and spice tones dominate these regions and there are many decorative techniques such as resist dyeing (batik) and tye dyeing using natural plant, flower and vegetable dyes. Typical colors are reds, yellows, dark greens, frequently decorated as a secondary application of metallic and wooden beads, gemstones and metallic inks. An annual design competition called MBOISA (most beautiful objects in South Africa) showcases locally designed clothing, furniture, housewares and jewelry. Gavin Rajah was a featured clothing designer whose Pebble Dress was a simple gold dress that was textured in a pattern of large smooth stones that are common in riverbeds in South Africa. Mikahla Hawker of Milky Rose Design used found objects to create votive candle holders in shades of brown and gold mesh. Sandy Godwin created a bone China set that were fired in a bright white clay, overlaid with gold flowers and leaves that are indigenous to South Africa. The tea set itself is also a nod to the colonialism era that forever shaped the culture of South Africa.

India

The warmer, sun-ridden climate of India has allowed for a bright subtropical pallet to endure. Bright colors (dyes) have been used for centuries since the early natural dyes would fade in sunlight so garments and textiles would initially appear quite vibrant and then gradually fade when bleached by the intense sun. These natural dyes were replaced by their synthetic counterparts in the early part of the 20th century but the bright, unfaded colors remained as the center of their shades.

Color Trends and Selection for Product Design.
DOI: http://dx.doi.org/10.1016/B978-0-323-39395-9.00022-0

The traditional Indian color pallet contains bright pinks and purples, navy blue, white, bright yellows and greens. Indians celebrate the Holi festival by sprinkling brightly colored powders all over each other as an homage to their Buddhist deities.

North America

United States and Canadian pallets tend to mirror traditional European pallets; however, Native American and Western and South Western US pallets are quite different. This pallet is associated with desert flora, fauna and minerals that have a subdued range of earth tones punctuated with vibrant turquoise, stark whites and clean blue shades. Sunset colors of pinks, yellows and oranges also compliment this pallet.

Latin America

Mexico, Peru and Argentina have pronounced earth tone pallets similar to the Native American pallets of the United States. The primary difference is the use of black and dark browns as contrast colors rather than white. These pallets also are brightened with pink and peach tones.

The mountainous country of Guatemala is the exception to these earth tones and its pallet stands out with its vibrant shades of reds, pinks purples and bright greens and oranges. These are the colors of the foliage of the Guatemalan rain forests and the vestiges of the Mayan culture that once dominated the region.

Further Reading

http://www.redbubble.com/groups/true-colors-of-asia/gallery
http://www.shikenan.com/blog/value-of-color-and-symbols-in-african-clothing/
http://www.thecolorsofindia.com/
http://www.holifestival.org/
http://www.warpaths2peacepipes.com/native-american-symbols/color-meanings-symbolism.htm
http://www.webdesignerdepot.com/2012/06/color-and-cultural-design-considerations/

23 Color Selection for Environments by Industry

Healthcare

Healthcare designers tend to cite the Purkinje effect that was discovered by a Czech polymath in 1819 who took meditative walks in the Poppy fields of Bohemia. He noticed that the poppies appeared very bright in the afternoon sun but reflected an almost mordant shade of brown or black at dusk. This effect states that in areas of lower illumination, reds and yellows tend to darken and disappear whereas blues and greens become more distinct. Color is frequently used to separate floors from walls for the visually impaired and elderly. Brighter colors for the elderly are frequently used in institutions since older patients tend to prefer colors that are easier to see and separate from each other.

Light colors are also used to promote healing. Blue is a calming color for all ages but geriatrics tend to favor reds and greens as they age. White and yellow are the least favorites. Young children prefer brighter colors and were found to stimulate activity but shifted towards warm blues and greens as they matured. Colors found in nature like blues and greens punctuated with woods and darker mineral and stone shades are associated with healing. Frosted and textured glass was preferred to transparent glass since it provides more privacy and rest. Since many hospital patients are bed ridden and confined to a small space, color fatigue can set in rapidly. The use of muted wall colors across from a window can reduce eye strain or headaches. Men tend to dislike reds and purples in healthcare settings. However, red lights are sometimes used in hospital settings so patients can rest but healthcare attendants can read charts and adjust equipment without disturbing the patients at night. Color variation is also important to reduce the institutional effect of space. Using contrasting color, entrance space or an accent wall is a good method to counteract this institutional effect along with a variety of textures and natural finishes whenever possible. Blue-green walls in an operating theater tend to counteract the after-image of blood when the surgical staff need to rest their eyes.

Color Trends and Selection for Product Design.
DOI: http://dx.doi.org/10.1016/B978-0-323-39395-9.00023-2

Military Colors

In low-light situations, such as in control rooms of submarines or in aircraft, red-light illumination is frequently employed so that pilots can read the control panel instruments but still be able to see outside into the darkness of the ocean or the night sky. Military uniform colors tend to be strong and dark in shades of grays, blues and greens. These colors are offset with blood colors of reds, purples and blues and are frequently accented with gold and silver trim to indicate the supremacy of the forces.

Color Laboratories

Color chemists and those who develop, measure and evaluate colors need an environment with tightly controlled illumination. Generally speaking, hallways and overhead lighting is kept to a minimum and daylight illumination is essential as a measurement tool but also needs to be controlled. Although most color observation and measurement occur in a light booth or light box, colored and highly reflective lights, walls and surfaces should be avoided to incorrectly reflect on the observed materials.

Noise Colors

As noise pollution to continues to increase with the urbanization of culture, noise filtering or masking is becoming more important to enable people to sleep, think and work more efficiently. White noise was probably the first color to be identified and was used in doctors and therapists offices to provide privacy between patients in the office and those in the waiting room. These little white noise machines provide a full spectrum sound that can mask conversations or external street noise. The sound is similar to static from an untuned FM radio and has been used to help people relax. The reason the white "color" was chosen was because, like white light, the sonic frequencies are unfiltered, similar to the visible spectrum where white light transmits all colors unless filtered by a prism. If compared to the colors of the rainbow, red colors have the lowest frequencies and blue colors have the highest frequencies. Dark Red or Brown noise is the lowest color noise (currently) and has a low, rumbling quality to it that is somewhat random and has a distant relationship to Brownian Motion and its random movement of particles through a liquid or gas. Pink noise is a little bit higher in frequency and is similar to the constant sound inside a passenger jet. The opposite of Pink noise is Blue noise. This is a much higher-frequency noise that is quite unpleasant but is quite effective in blocking out unwanted, lower frequencies. It sounds more like a hiss. Most sound machines and apps offer White, Pink and Brown noise but leave out the Blue noise and add in multitonal combinations of waterfalls, rain, birds, etc. to help people relax or focus.

Further Reading

http://www.ki.com/pdfs/Color_in_Healthcare_Settings_Paper.pdf
http://www.ee.columbia.edu/~dpwe/noise/
https://en.wikipedia.org/wiki/Purkinje_effect
https://en.wikipedia.org/wiki/Brownian_motion

24 The Color Influencers

For those who create new colorants, there are a few organizations where designers, educators and color chemists meet to discuss the creation of new colors and the decline of others. Generally speaking, new colors are cleaner and brighter shades of their ancestors. These are developed by manufacturers to compete with their counterparts to increase market share. A typical salesperson will visit a paint manufacturer or plastic molder and will show them color cards of old versus new shades of their own colors or new shades versus the competitor's products. There are a number of features that manufacturers try to meet when making a new colorant including higher chroma (L value), higher hiding, more environmentally friendly (greener colorants) and unique areas of color space that were not possible with existing products.

Color Organizations and Industries

Color organizations and industries provide credibility and consistency to the color trend and popularity industries as well as vet and develop new technologies to measure and characterize colors across the globe and help to provide color harmony and consistency within specific industries.

Color Matching/Management Organizations

There are numerous matching and color systems currently available throughout the world. Some of these are based on their own color measurement theories and others are just a system of classification used to match colors for specific applications or industries.

Natural Color System

The Natural Color System (NCS) was developed by the Scandinavian Institute in Stockholm, Sweden that dates back as far as

Color Trends and Selection for Product Design.
DOI: http://dx.doi.org/10.1016/B978-0-323-39395-9.00024-4

1611 when the first recorded color theories were developed by Aron Forsius. The Scandinavian Institute was officially formed in 1945 to develop a color classification system. The first edition of the NCS was launched in 1979 with 1412 colors and has expanded to digital and cloud versions in 2014.

Toyo Ink

Toyo is a Japanese ink company that also produces a series of standardized colors. They published their first fan deck for inks in 1908 and were the first manufacturer to create a standardized pallet of inks. In 1995, they published their Worldwide 5-Region Network System of standardized colors.

RAL Matching System

The RAL German color-matching system (*Reichs-Ausschuß für Lieferbedingungen und Gütesicherung*) was developed in 1927 to standardize color selection for manufacturers and initially consisted of 40 color swatches for coatings and inks and later for powder coatings. The current classic system consists of 210 colors but they publish many different books or color fans that allow for different levels of gloss, the use of effect or metallic pigments and also a software version. RAL also publishes a year inspiration book identifying the main color trends that they are predicting for the next 2 years.

Munsell Color System

The Munsell Color System is a three dimensional color system that was developed by Albert Munsell, an American artist and Professor, who wanted to develop a rational method to describe color that used color numbers instead of names. He outlined three important components of color: hue, chroma and value (lightness). The hue is the shade of color and is divided into red, yellow, green, blue and purple. The chroma is a measurement of the saturation or pureness of the color. Lower values are less saturated (pastels) and higher values are brighter colors but there is no upper numerical limit for chroma. Lightness is also known as the "L" value where black is zero and pure white is 100. In color space, this system is represented as a sphere with a vertical axis running though the center of the sphere. The L value (lightness) is represented by the vertical axis and the chroma and hue comprise the horizontal component that creates the circular nature of the sphere. In 1930, this system was adopted by the USDA as the official color system for soil classification.

Pantone, Inc.

Pantone is a color direction and publishing segment of the X-Rite Corporation based in the United States. They are best known their matching system which are books of thousands of standardized color swatches that are composed of pigment mixtures in paper, plastic or fabric. Each color is given a Pantone Matching System reference number, and in certain cases, a color name and spectral data such as RGB or LAB references. These books are frequently used by designers who select their desired color for a project and send the chip or reference number to the manufacturer to create in the paint, plastic or fabric that will be used for their project. In addition to their swatch books, digital files are also available but are much more limited in their matching ability due to the lack of uniform calibration from one display monitor to the next as well as their inability to accurately reproduce luster, sparkle, reflectivity or color travel from effect pigments. Pantone also nominates a "Color of the Year" based on their own internal research.

CIE: International Commission on Illumination

The International Commission on Illumination is also known as the Commission internationale de l'éclairage (CIE) was founded in 1913 and is the international authority for its command of color, illumination and color space based in Vienna Austria. In 1931, CIE developed a method to determine the three dimensional color space of an object based on wavelengths in nanometers in an XYZ plane. The CIE XYZ color space includes all color sensations that an average person can see. This is a standard reference against which many other color spaces are defined. A set of color-matching functions, like the spectral sensitivity curves of the LMS space (long, medium and short wavelengths) but not restricted to be nonnegative sensitivities, associates physically produced light spectra with specific tristimulus values. The tristimulus values spaces are the three primary colors (RGB) in an additive color model.

A typical human has three different types of cone cells in the eye. The cone cells are responsible for detecting different wavelengths of light. Another type of photoreceptor cell in the human eye is the rod cells which are responsible for detecting objects in low-light situations.

Most humans have S-Cones (short wavelength) blue receptors, M-Cones (medium wavelength) yellow/green receptors and L-Cones (long wavelength) red receptors. This is known as Trichromatic Color Vision. There have been some documented findings of humans with four types of cones that have a higher sense of color discrimination known as tetrachromats. Most trichromats can see 1 million different hues and most tetrachromats can see in the range of 100 million hues. Many birds and other living things are also tetrachromats.

ICC: International Color Consortium

The International Color Consortium was founded in 1993 by eight different vendors who were interested in creating color parity between digital devices using a color management system. A color management system is one that translates digital data from one device to another. An example of this would be a camera or scanner that captures an image in RGB and then uses an ICC profile to translate that data into CMYK to a printer that prints out those colors. This profile consist of tables of data that relate the device coordinates to those of the standard color space defined by ICC based on a measurement system defined internationally by CIE. This will provide a relatively accurate representation of the color perceived by one device (such as a camera) that will be detected and printed as a relative color image (such as printer).

AIC: International Color Association

The International Color Association or the Association of Internationale de la Couleur is a learned organization that had its genesis in 1957 with the French *Centre d'Infomation de la Couleur* (CIC) when they published their first international journal of color. The organization grew to include other countries throughout Europe and published their international journal of color in France every 2 years and in a different member country in rotation every year in between. The AIC foundation officially occurred in 1967 in Washington DC, United States, during the 16th Session of the CIE (Commission Internationale de l'Éclairage). This organization encourages research and dissemination of information in all areas of color including design, art, science and industry on an international scale. The other major tenet of this organization is to cooperate and not compete with other color organizations including ISO (International Standards Organization), CIE (International Commission on Illumination) and ICO (International

Commission for Optics). In 2008, they also designated March 21st as the International Color day since the days around this date have an equal amount of day and night (equinox).

Further Reading

http://www.ncscolour.com/en/about-us/our-story/
http://www.colorassociation.com/pages/2-about
http://www.aic-color.org/index.htm
http://color.org/faqs.xalter
http://munsell.com/
https://en.wikipedia.org/wiki/Pantone
https://en.wikipedia.org/wiki/RAL_colour_standard
http://www.toyoink.com/about/history/

25 Color Forecasting Trade Associations

Color forecasting trade associations are essential for manufacturers of different industries to come together to discuss past, present and future colors. Most brand owners use this as a sounding board to ensure that they do not make fatal errors when deciding on colors for their products.

Errors in Color Selection

As an example, hospitals frequently use color-coded wristbands to signal the needs and special requests of patients. Some hospitals have reported that using yellow wristbands for do not resuscitate (DNR) patients has resulted in some near misses where patients who were wearing yellow "Live Strong" bracelets were mistaken for DNR wristbands. Some states have standardized the color of wristbands between hospitals to indicate different treatment modalities but the standardization varies from state to state. Endurid is a company that manufactures wristbands for patients and inmates, has created an international standard for five color codes for wristbands. Red is indicated for allergies, yellow is a fall risk. Rather than using yellow for DNR, they have chosen a periwinkle blue. Green is the color for latex allergy and pink is the color for restricted extremity. The restricted extremity band is placed on the affected arm of the patient that may be injured or swollen to indicate that arm should not be used for a blood pressure cuff, blood draw or any other type of analysis or monitoring.

Another example of a poor color choice is when a detergent manufacturer chose to color their liquid detergent bottles black. The rationale was to indicate that the detergent was a very powerful cleaning agent as well as a premium product with a higher price point. The reality was that the public perception was that the product was dirty and ineffective as a detergent. The color is no longer used for that detergent. Another poor example is when red is used as a decorative color for razors, toothbrush bristles or feminine protection products. Red is quickly associated

Color Trends and Selection for Product Design.
DOI: http://dx.doi.org/10.1016/B978-0-323-39395-9.00025-6

with blood and all of these products are looking to avoid or protect from bleeding rather than enhancing the relationship. In a New York City museum, the men's restroom sign has a pink background and the women's restroom sign has a silver background. The misinterpretation is obvious due to a poor color choice. The United States military had to discontinue the use of rectangular-shaped yellow food bags that were being air-dropped in Afghanistan because they were being confused with the cylinder-shaped yellow cluster bombs the United States had been dropping on the same areas earlier.

CAUS: Color Association of the United States

This organization is based in New York and was founded in 1915 as the Textile Color Card Association (TCCA) as a response to the burgeoning Industrial Revolution in the United States and because less information about colors and trends was available from Paris since devastation of the two world wars of the 20th century. The millinery industry, including hosiery, hats and gloves created a standardized set of colors for the textile industry in the form of a color card of 106 colors that were distilled from a selection of over 2000 colors that were being used by the industry at that time. The 106 colors that were chosen (all by men) were stated to represent 75% of the total color consumption of the textile industry. Each color was given a key that indicated its hue, level of chroma and any blends of color that would comprise this shade. The idea was to distribute this card to colorant manufacturers, textile dyers and weavers, and retailers to create a standard for uniform colors across their industry. Currently, this for-profit organization provides very sophisticated guidance for educators and businesses that are looking to provide global color intelligence for branding and products based on numerous criteria including gender, cultural, age, psychological demographics, market segments and trends rather than the standardization of color shades.

ICA: International Color Authority

The ICA is a privately held business that creates color trends for future time periods and seasons, used by fashion, textile and industrial designers. These forecasts were first published in London in 1968. A panel of international color gurus meets every 6 months to forecast the new colors. These colors are predicted about 2 years ahead of time. Their forecasts and color guides are published and directly referenced to Pantone and NCS color standards to provide uniformity in colors across different industries and methods of conveyance.

CMG: Color Marketing Group

Color Marketing Group is a US-based organization that was founded in 1962 and consists of color professionals from all over the world that meet on a regular basis (usually two times per year) to determine the existing, emerging and future colors for different industries for that year. Membership is based on recommendation and professional color experience.

Further Reading

http://color.org/faqs.xalter
http://endurid.com/2013/06/17/the-5-different-colors-of-medical-alert-bracelets/
http://query.nytimes.com/mem/archive-free/pdf?res=9B0CE3DF1038E633A25754C0A9609C946496D6CF standard color card of America
http://dbpedia.org/page/International_Colour_Authority
http://www.catherinemaustin.com/tag/international-color-authority/
http://www.colormarketing.org/
http://www.colorassociation.com/pages/2-about
https://en.wikipedia.org/wiki/International_Colour_Day

26 Color Events

There are numerous events and venues that influence color trends globally. Some of these are actual events that consumers may visit and some are virtual and can be viewed on-line.

DIY/Innovation

We are currently living in an age of do it yourself (DIY) popularity. This trend is due, in part, to economic hardship and in part due to a backlash of consumers inundated with disposable products that are simply discarded when they are depleted or fall into a state of disrepair. This has inspired consumers to reuse, repurpose and fix items that are no longer useful. It has also spurred consumers to create simpler products that work better or less expensively for their own needs. DIY has also been fueled by society's lack of practical education. As employers require workers with highly specialized job requirements, schools and universities have evolved their studies to meet these needs. These specialized courses of studies have eliminated more practical curricula and students graduate with a lack of basic life skills and knowledge. Basic knowledge about how to make or fix things are a very small part of formal education and consumers look to DIY websites and events to learn about and display their projects.

Maker Faires

The Maker Faires are local events that are held around the globe for average consumers to show off their inventions and collaborate with other like-minded inventors who have similar ideas and designs or for consumers, designers and engineers who are just looking to be inspired by other people's ideas. Some of these are very basic ideas like constructing wooden toys and weaving fabrics. Other exhibitors use incremental innovation to create new things from currently existing technology. Many of these projects use open-source innovation so that they can freely collaborate with others to refine their projects. The color

Color Trends and Selection for Product Design.
DOI: http://dx.doi.org/10.1016/B978-0-323-39395-9.00026-8

pallets vary widely depending on the application. Many of the hands-on types of projects feature very natural wood and metal colors. Many of the 3D printing applications feature equipment that are also quite natural (raw-wood colors) or low-tech blacks and silvers. However, filament colors are quite saturated and bright.

Open-source innovation is a relatively new trend that developed from the 1960s idea of open innovation where companies leased or purchased technologies from other companies. Open-source innovation became popular in the late 1990s when software companies decided to release some of their products to the public domain. This concept has evolved over time to include innovations and intellectual property that is released or donated for charitable causes. It has also evolved into an antipatenting trend that is becoming common-place with many companies who realize that they need to look beyond their own R&D teams and think-tanks to keep up with their competitors and realize the cost-prohibitive/indefensible nature of the patenting process across international boundaries.

Fashion Shows (Runway Events)

Fashion weeks are a large-scale event where many clothing designers show their new collections to buyers and the media several months before the start of the season so that they can take orders and manufacture the garments in time for the season to start. Most cities in the world sponsor a fashion week or weekend. The four major fashion week shows occur in New York, Milan, Paris and London twice per year. The major shows are Spring (Spring/Summer) & Fall (Autum/Winter) Ready to Wear, Spring & Fall Menswear, Spring & Fall Bridal, Resort wear (before Spring/Summer) and Pre-Fall (before Autumn/Winter), Spring & Fall Couture and Pre Fall.

There are also different genres of fashion week including Christian Fashion Week. CFW started in February 2013 in Tampa, Florida and is an annual fashion show that celebrates fashion from a Christian faith perspective and features modest clothing. The first event attracted several hundred attendees and several thousand virtual viewers. Some of the fashions that were featured at the 2014 show were Pro-Israel handbags with doves of peace attached. Glittery black t-shirts with bible verses prominently displayed and Christian swimwear that featured modest skirts and coverups. The Bangalore Fashion Week in India features festive wear. BFW features very traditional Indian clothing that consists of long, diaphanous fashions, rich colors and metallics and frequent use of Nehru collars.

The Portland Fashion Week in Portland Oregon started in 2013 and features green and sustainable clothing using eco-sustainable manufacturing.

New York

New York City is the primary city for fashion shows in the United States. The two fashion weeks in New York occur in September and February and hosted the first fashion week in the world. It started in 1943 and was called Press Week. This event started, in part, because the press was unable to travel to Paris because of World War II. Another reason for this event was to promote the American fashion industry as separate and distinct from the French industry. Over the years, NY Fashion Week has been held in different areas of New York City including Bryant Park and Lincoln Center. In 2007, Mercedes Benz became the primary sponsor for NY Fashion Week and the event was called Mercedes Benz fashion week. In January 2015, Mercedes Benz announced that they will no longer be the title sponsor of the event but that they will continue to sponsor events in other cities around the world. The name of the event will revert back to New York Fashion Week and the new location of the event has not yet been revealed. The Spring 2016 RTW showed many fashions styled in black and white. Reds were more natural and shades of brick and russet were prominent. Dior launched a new miniduffel bag in shades of black, white, hunter green and bright blue. The overall show featured very bright almost rainbow hues.

Paris

Couture Paris fashion shows were started in 1945. Paris Fashion week was started in 1973.

The Couture shows take place in Paris in January and July and are restricted to Paris by the Chambre Syndicale de la Haute Couture. The current location is in the Carrousel de Louvre. In 2015, the French Ministry of Culture featured their first Dwarf runway show during the fall fashion week in Paris. The Spring 2016 show featured a profusion of light blue colors in shades of robin's egg, light turquoise and powdery blues. Chanel featured a Birkenstock-type of light up, black sandal with bright blue crystals worn with white socks. The Valentino show featured many dark and somber colors that transformed into some mustard yellows and metallics.

Milan

The Milan Fashion Week started in 1958 and the major fashion shows are the Milano Moda Donna and the Ready to Wear Spring & Summer event. Light pink colors dominated the runways of the Milan Spring Summer 2016 show. Max Mara showed long, flowing fashions in delicate pink and taupe shades. Moschino showed knitwear in a study of lemon and acid yellow shades, bright oranges and acid greens. Jil Sander showed pale and delicate tones of pink and white contrasted with black and dark blues. Another color scheme that was trending was dark, almost deep sea shades of blues, blacks, purples, punctuated with bright corals and electric blues.

London

The London Fashion week was started by the British Fashion Council in 1984 by bringing a disjointed set of London designers together to create a cohesive show featuring British fashions. In 1986, Lady Diana visited the show and created the first formal link between London Fashion Week and the British Monarchy. Margaret Thatcher also attended the show. Alexander McQueen debuted his fashions in 1992. In 2002, Anna Wintour attended her first London Fashion week to show her gratitude since the London show adjusted their dates to accommodate the New York Fashion Week date change due to the September 11th attacks in New York. The fashions from the Spring/Summer 2016 RTW line included lots of folded and ruffled, very feminine fashions in mostly in tone on tone whites, tiny floral prints and deep purples. Many designers combined these frilly flounces with large, bold black and white geometric patterns. Martin Margiella showed an extraordinarily pale blue bustier with a silver metallic bra that was pared with flowing off-white slacks, pure white blousey arm bands and black and silver platform shoes.

Tokyo

Mercedes Benz Fashion Week in Tokyo is the most important fashion event in Japan. In 2015, MBFWT is sponsoring Tokyoeye 2016 where a handful of small and medium RTW women's designers will be featured. Motohiro Tanji is a designer of stunning knitwear for women. Many of the dresses and sweaters were gargantuan entanglements of fibers in whitened tones, dark grays and blacks and midnight blues. Motonari Ono show exhibited styles that were more costume than fashion and had a definite 1950s/1960s flavor with pillbox hats, bright red lipsticks, leopard print tight slacks, excessive tartan and large plaid ensembles that resembled a Burberry disruption. His color schemes were black and whites, accented with reds. Goldenrod yellows and combinations of blacks, browns and taupes.

Red Carpet Events

There are many different types of Red Carpet events but for the sake of simplicity, only a few of the more significant Red Carpet events will be discussed that help shape color trends for many different industries, including the EGOTs (Emmy, Grammy Tony and Oscar) award ceremonies.

CFDA Awards

This award ceremony is sponsored by the Council for Fashion Designers of America. It is frequently referred to as the Oscars of fashion and honors excellence in fashion design. The organization was founded in 1962 to promote the American fashion industry throughout the rest of the world. However, as we have emerged into a more global economy, many of the awards recipients have been from France, Italy, Japan and the United Kingdom. Colors and fashion styles were all over the lot at 2015 event. As with most of these red carpet venues, there is a preponderance of black. However, the red shades were more to the yellow and bordered on coral and orange shades rather than the traditional dramatic blood red shades.

Met Gala Red Carpet

The Met Ball is an annual fundraising event for the Costume Institute at the Metropolitan Museum of Art in New York City. This event started in 1972 and every year has a different theme that relates to the exhibition that is opening at the museum that year. Each guest is expected to dress in clothing that is inspired by that theme. As an example, in 2015 the theme was "China: Through the Looking Glass." This exhibition explores the influence of Chinese culture into past and present fashion. The looking glass refers to how fashion designers were inspired by Chinese culture and implemented that culture into their fashions. A good example is Alexander McQueen who loved the blue and white porcelain that is common in Chinese design. McQueen took pieces of porcelain, smashed them into small pieces and sewed them onto a dress. As expected, many of the guests arrived in fashions that were gold in color or adorned in gold. Red was the color that followed most closely to gold followed by emerald and jade shades of green.

Tony Awards

The Tony Awards ceremony for theatrical productions that started in 1947 in New York City. This event is held annually in New York and is considered to be one of the least fashion-inspired events of the EGOT events until 2015 when Anna Wintour (Editor in Chief of Vogue) was asked to heighten the fashion sensibility of the ceremony. Reds and russets dominated the fashions and closely followed by sequined blacks and whites.

Oscars

The Academy awards or Oscar Awards in Hollywood, California are the most televised events of all the EGOTs with live feeds in all US time zones (except Hawaii) and also in the United Kingdom and Canada. The purpose of this event is to present those who create and participate in the film with a gold statuette, commonly referred to as "Oscar." This is a black tie event for men and women are expected to wear fashions created by the top designers of that year. There are several events that happen before and after the actual Academy Awards event. Most notable are the Vanity Fair After-Party and the Governor's Ball.

The Vanity Fair Oscar After-Party is a fashion forward event that enables many attendees to wear a second outfit for the evening.

Emmys

The Emmys are a category of entertainment awards for the television industry. The first awards show was held in Los Angeles in 1949 and was designed as a regional event for shows created in the Los Angeles area. The awards are subdivided into six categories including Daytime, Primetime, Sports, News & Documentary, International and Engineering

Awards. The name "Emmy" comes from the statuette that was created for the award that is given to the honorees. Originally the award was called an "Immy" that was a common name of the image orthicon tube that was part of the camera technology in the 1940s and 1950s. However, the name was changed to a more feminized version: Emmy that corresponded to the statuette on the award. There are numerous receptions and celebrations that occur throughout the year that provide opportunities for the media to review and analyze the fashions that are displayed on the Red Carpet events. Everyone seems to have an opinion on the fashions that are on display on these events including The Fashion Police from the Entertainment Television Network, The Wall Street Journal, The New York Times, The Chicago Tribune, The UK Guardian, etc. Generally in the United States, all of the morning network syndicated news programs feature a spot on the fashion hits and failures of these events. All of this media attention shapes the opinions of the viewers as well as the career paths of the designers whose fashions are reviewed. As expected, reds, blacks, whites and metallics favor these events but occasionally a unique color stands out either as embracing the current color trend pallet or seriously misaligned from the times.

MTV Video Music Awards (VMA)

Prior to the 1980s most popular music was heard and not seen. Most music celebrities were rarely seen in concerts, television spots or public appearances when compared to the massive exposure that they received from their music videos created for television that were launched on the Music Television Video channel in 1981. This constant exposure changed the faces, voices and fashions of music celebrities forever. It is true that the Beetles used music videos to promote their art prior to the launch of MTV but most of their contemporaries had much less media exposure. The Beetles influence in fashion and culture can be, in part, attributed to their mass-exposure from these television appearances. There were other music television shows (not networks) that were also influential in music, fashion and dance including the Dick Clark's show, American Bandstand, The Lawrence Welk Show, Soul Train, Top of the Pops (BBC) and Austin City Limits. The MTV music video and fashion awards are an important part of the Red Carpet events that influence fashion and color trends throughout the world. The VMAs were initially conceived to be an alternative to the Grammys. The first VMA was held in 1984 in New York City and the statue is an astronaut on the surface of the moon holding an MTV flag. The fashions of the

VMAs are more costume than couture but are important signals to future trends in serious fashions and colors.

CMA

The Country Music Awards are sponsored by the Academy of Country Music. The ACM was founded in 1964 to promote American country music in the 13 western states and the west coast in the United States. The first Country Music Awards ceremony was launched in 1966 in Los Angeles, CA. Country music has long promoted more casual fashions and their adjacent color stories than what is typically featured by other types of popular and classical musicians and cultures. Wrangler blue jeans and dark colored plaid or embroidered shirts with pearlized snap closures types of western wear with cowboy boots and hats have been historical fashion staples of the Country Music uniform but this has been changing. The western snap shirts have been replaced by simpler embroidered or brightly blocked t-shirts. In addition, many of the country music celebrities favor statement t-shirts that promote their causes and beliefs. Additionally, the pristine Wrangler jeans have been replaced by torn and highly embellished and embroidered blue jeans. In the 2015 awards ceremony, black leather, black denim, black tuxedo jackets over casual shirts and black rhinestone-embellished dresses dominated the event. As a contrast, Carrie Underwood appeared in a long white silver and gold-fringed dress in 2014. In 2000 Dolly Parton appeared at the awards in a pink leopard, sequined dress. In 2011, Ira Deane appeared a Pucci-inspired swirling suit in gelato shades of tangerine, chocolate and raspberry.

Obies

The Obie awards are defined as the off-Broadway theater awards given annually in New York City. The awards were initially given by the Village Voice newspaper to distinguished actors and groups that performed in off-Broadway theater productions. In 2014, the American Theater wing jointly presented the awards with the Village Voice. The fashions and the colors of the event are atypical. Celebrities frequently combine couture with casual fashions, sport unique eyewear and cross-dress. All of these statements send subtle messages to the fashion community and society at large about the upcoming trends in color and design.

Further Reading

http://www.examiner.com/article/46th-annual-academy-of-country-music-awards-worst-dressed
http://stylecaster.com/governors-ball-street-style-2015/%20-%20autoplay
http://www.nytimes.com/2015/05/08/arts/design/review-in-china-through-the-looking-glass-eastern-culture-meets-western-fashion.html?_r=0
http://www.emmys.com/
http://www.clothes2order.com/read_news/69/4/The_Changing_Trends_of_Country_Music_Clothing.html
https://en.wikipedia.org/wiki/Tony_Award
http://cfda.com/about
http://www.apartmenttherapy.com/whats-new-whats-next-5-hot-decor-trends-straight-from-the-runway-219685
http://fashion.telegraph.co.uk/news-features/TMG6194975/25-years-of-London-Fashion-Week.html
http://www.buzzfeed.com/jessicamisener/61-things-i-saw-at-christian-fashion-week#.pgyJdMpQ5
http://nymag.com/thecut/runway/
http://www.foxnews.com/entertainment/2015/10/03/french-ministry-hosts-dwarf-fashion-show/
http://nyfw.com/
http://www.oregonlive.com/shoporegon/story.ssf/fashion/portland-fashion-week-oct-7-11.html
http://www.vogue.co.uk/fashion/trends/2016-spring-summer/london-frills-and-flounce-trend/gallery/1476530
http://www.sitepoint.com/10-troublesome-colors-to-avoid-in-your-advertising/
http://www.sciencedaily.com/releases/2007/12/071218122409.htm
https://en.wikipedia.org/wiki/Open_innovation
http://endurid.com/2013/06/17/the-5-different-colors-of-medical-alert-bracelets/
https://www.pinterest.com/pin/377598749984979058/

27 Art Shows and Fairs

Art Shows, Fairs and museum installations (to a lesser extent) are many times regarded as the birthplace for color trend and design movements. The emerging artists or newest movements are, arguably, the most important segment of these venues. Museum shows and installations are usually compiled over several years or decades (The Whitney Biennial) of an artist's work and strive to represent comprehensive distillations of past and current trends. Most Art Fairs are a local collection of galleries that exhibit their most prominent and emerging artists to attract the attention of the art collectors and buyers. These shows are also frequented by industrial designers, architects, students and other artists looking for inspiration for their own work.

New York Armory Show

This show was initially held in 1913 in New York City and was regarded as one of the most influential art fairs in the world for modern and contemporary art. The original venue was the New York City 69th regiment National Guard Armory in Manhattan. This site was chosen, in part, due to its vast interior space as well as its centralized location in New York City. The Armory Show went on tour and was set up at the Art Institute in Chicago and the Copley Center for Art in Boston. These types of European modern art movements such as Fauvism and Cubism were virtually unknown to Americans and served as inspiration to American artists to develop their own forms of modern art in the coming years. Currently, the Armory Show is held every March in New York City on Piers 92 and 94. In 2015, the top seller of the show was a piece by Lucio Fontano entitled *Concetto Spaziale, Attese* (1966) which is a warm red monochrome with seven tagli sliced across the surface, creating a rippled effect. The average selling price for art at this show is between 20,000 and 100,000 USD. This show is the pivotal venue for the Armory Arts Week that showcases nonprofit art throughout New York during the same week in March.

Color Trends and Selection for Product Design.
DOI: http://dx.doi.org/10.1016/B978-0-323-39395-9.00027-X

Art Basel

Art Basel is commonly referred to as the "Art Olympics" of the international art world for modern and contemporary art that allows galleries to showcase their best art to global collectors, museums and curators. Basel in Switzerland was the birthplace of this art and cultural event in 1970 by three local galleries in Basel to showcase the artists that they represented. In 3 years, they had several hundred exhibitors from over all over the world. The parent company is MCH which is a Swiss holding company that organizes tradeshows and international exhibitions including many jewelry and watch shows. This indicates the fine line of influence that runs between the art and commercial worlds of fashion, jewelry and technologies and how they influence each other and inspire emerging trends.

In June 2015, Art Basel featured the art of Korea in their Feature Korea exhibition. This is sponsored by the Korea Ministry of Culture, Sports and Tourism and showcases contemporary art trends that are current in Korea. The art pieces were quite diverse from an outdoor heart sculpture in tones of gray to an enormous textured wall hanging that created and undulating black effect. Another exhibition that is a staple of the Art Basel show is the Breeder program that features emerging artists and has long been recognized as an incubator for new expression. The Breeder program was started in 2001.

Art Basel Miami

In 2002, MCH launched the Art Basel fair in Miami which is now almost as large as the fair in Basel, Switzerland. Art Basel coincides with the Miami Art Week, the first week in December.

Art Basel Hong Kong

In 2011, MCH acquired controlling interest in Asian Art Fairs Ltd. And Art Basel Hong Kong was opened in 2013 and was the first international, large-scale fair of contemporary and modern art in Asia.

Further Reading

https://en.wikipedia.org/wiki/Art_Basel
https://scope-art.com/shows/basel-2015/about/
http://www.huffingtonpost.com/2014/03/06/whitney-biennial-2014_n_4897931.html
http://www.thearmoryshow.com/
https://en.wikipedia.org/wiki/Nude_Descending_a_Staircase,_No._2
https://en.wikipedia.org/wiki/Armory_Show#cite_note-Brown.2C_Milton_W._1963.2C_pp._185-2
https://scope-art.com/shows/basel-2015/about/
http://www.motohirotanji.com/

28 Industry Trade Shows

Every day, there are hundreds of trade shows and fairs all over the world. The exhibitors are usually the leaders or the emerging leaders of these markets and the visitors are usually the customers and decision makers for these markets and products. Invariably, there is media presence at these events to analyze and report on these venues. It is impossible to cover every type of tradeshow but the most color-influenced venues will be highlighted.

Toy Fairs

Toy shows are one of the best ways to track technology trends as well as excellent segmentation for color popularity for infants, toddlers, children, preteens and teens. Two of the most important shows are the New York Toy Fair in New York City and the International Toy Fair in Nuremburg, Germany.

New York Toy Fair

The New York Toy Fair started in New York City in 1903 a few weeks before Christmas. It is the oldest venue of its kind in the world and the first show featured about 10 manufacturers, including Lionel, the classic toy train manufacturer. According to the Toy Association of America, the primary themes for 2015 were "Maker Movement," "Open-ended Playtime," "Mini Madness," "Smart Play," "Under the Sea," "Top in Tech" and "Dawn of the Dinosaurs." "Maker Movement" toys focus on creative building and inventions. The color pallets are bold primary colors of red, blue and yellow. This is a category that favors boys of all ages. "Open-ended Playtime" toys include toys that enhance creativity and imagination with products such as modeling clay, blocks, crayons, dolls, etc. These colors are more whimsical with shades of purples, pinks, rainbow pallets and outer space high-tech colors. This is more of a preteen girl pallet. "Smart Play" toys are classic toys and STEAM (science, toys, engineering art and math)

Color Trends and Selection for Product Design.
DOI: http://dx.doi.org/10.1016/B978-0-323-39395-9.00028-1

enhancement. This includes books and board games that teach advanced math and coding and enhance cultural sensitivity. Color pallets are subdued earth tones and metallics that is also gender neutral but skews slightly more to preteen boys. "Mini Madness" is all about miniature toys that are sold at a lower price point and are ideal for lower income segments, travel toys as well as party favors. This segment includes small die cast cars, small dolls, guns and balls. This is also a gender-neutral pallet. Color pallets are over the top bright and fluorescent high-impact colors of oranges, reds, pinks and bright greens. This is a gender-neutral pallet. "Under the Sea" products are targeted primarily to baby and toddler bath and pool play and skews slightly to toddler girls. "Top in Tech" is a category that includes R/C toys and vehicles, toys that sync with apps and electronic wearables. The color pallet is mostly metallic, with subdued primaries with a few feminine pinks and purples. This is an older segment appealing mostly to preteen and teen boys. "Dawn of the Dinosaurs" is a category that includes plush and molded dinosaurs, books and games that enhance learning about dinosaurs and appeals mostly to toddler boys. The color pallet is dark greens, browns and blacks punctuated with oranges, reds and greens.

International Toy Fair, Nuremberg Germany

The Spielwarenmesse *International Toy Fair*, also known as "The International Toy Show" is the largest toy fair in the world. Although it is similar in size to the New York show, the number of international exhibitors and visitors far exceeds the New York Fair. The show is also newer than the New York show, starting in 1949. The city of Nuremberg has always been famous toys since the beginning of the dollmaker industry in medieval times as well as the invention of tin toys during the industrial revolution. An interesting observation about the American show that is different from the German show is the target market. The New York show and most of its products are geared towards children that are quite segmented whereas the German show tends to promote more toys and products for children of all ages, including adults. Traditionally, toys trains were designed and targeted towards children. However, in the past 50 years, these toy trains and other products such as Legos have resided in the domain of adults who have never let go of these toys. This trend is more prominent in the European show than the US show. This also points to a difference in quality of toys. Toys that are geared towards adults are generally more costly and are higher in quality. There is a marked difference in overall quality and price point of toys that originate at the two shows. One of the 2015 trends at the Nuremberg show was featuring toys in a category of "Design to Play," these are toys that function as playthings but are also esthetically pleasing enough to display in the living room or bedroom as decorative furnishings for the home. This segment includes basic toys such as model cars and animals made out of natural materials such as wood and tin. They tend to follow art movements such as Bauhaus and Pop Art and have natural finishes and colors. "Baby Boom" is a trend that focuses on toys for babies whose parents are older. Many of these products relate to travel and transportation, focusing on high quality and status prams, push chairs, car seats and diaper bags made from the finest materials with leather handles, sturdy and stylish fabrics such as vintage denims and highly targeted styles for gender differentiation. Also manufacturers are bundling multiple car seats and prams to allow for grandparents and multiparent families to share in childcare. The color pallet is geared primarily towards automotive colors, the exception being the highly specified gender segmented products that are loaded with pinks, purples, sparkles, yellow, red and blue construction vehicles, rockets and trains.

Another pronounced trend that exists at both shows is the creation of toys that relates to commercially licensed images and logos. A perfect example will be toys that will arrive in late 2015 and early 2016 that relate to the release of the new Star Wars movie. This is an interesting trend where product colors are directly influenced by entertainment events.

Furnishing and Interior Design Shows

Interior designs and furnishings are a market that turns over quickly where most consumers redecorate every 2−3 years so this is a market that pays very close attention to the fashion market because they also complement and share the use of textiles in their products.

Techtextil and Heimtextil

The Technical Textile (Techtexil) and the home textile show (Heimtexil) both occur in Frankfurt Germany on an annual basis. Both of these shows are very important for trend analysis because they cover two very different sides of the textile industry.

Heimtexil focuses on home and contract textiles and features a themepark where it investigates textile trends that feature patterns, finishes and colors. In 2015, the trends focused on the senses and were entitled

"Well-being" and covered such topics as protection (white story), energizing story (vibrant shades of reds, blues, yellows and pinks), a green healing story of neutrals and greens and a nouveau riche story of multiculturalism in a pallet of metallics.

Techtextil is a show that features technical textiles and nonwovens for different industries including Agritech (Agricultural), Buildtech (construction), Oekotech (eco), etc. Other techtextil shows are held in the United States, Russia and India every 2 years.

Industrial Contemporary Furniture Fair

ICFF is the Industrial Contemporary Furniture Fair in New York and is the pivot point for the annual New York Design week where many museums and design groups in different neighborhoods in New York sponsor design-related events and installations. Some of the trends evident in the 2015 show were classic designs that were reintroduced in novel colors and finishes, neutral landscapes with one brightly colored piece of furniture or accessory and brightly colored textiles in Rorschach and graffiti patterns.

Neo Con

Neo Con is the interior design and furnishings show held in Chicago's Merchandise Mart and is a pivotal showcase for flooring textiles, furniture and window treatments. It is the largest interior design show in North America and started in 1969. The 2015 Neo Con show exhibited many new and interesting designs. Bernhardts's Brandon Kim created the Lilt chair that has a triangular shape which makes the occupant feel as though they are sitting on a tree limb. This chair is featured in chartruse, white, turquoise and spruce green. The Andaz seating collection by Tod Bracher was designed for casual cafes with an array of stools and stackable chairs in a 1980s colorway gray, deep purple and beige fabrics with simple chrome legs. Mockett introduced a 3 outlet that is a recessed grommet mount called Flip that is available in black, white, orange and chartruse in a soft touch plastic finish.

Milan Furniture Fair

The Salone Internazionale del Mobile or Milan Furniture Fair is held annually in Milan, Italy and is the largest interior fashion show in the world. As with the ICFF, the Milan Furniture Fair is the pivotal event in Milan's design week with representatives from electronics, automotive and fashion. The fair started in 1961 by furniture manufacturers in Italy but has grown in size and international influence over the years. Some of the highlights and trends from the 2015 show were Louis Vitton's "Objet Nomades" a series of portable chairs and fold-up travel accessories designed to make rigours of travel more tolerable and refined. Raw leather colors were combined with deeply tanned and burnished textiles. Another trend that was evident was furnishings that were created in natural food colors of browns, greens, aubergine and oranges. This might be due, in part, to the fact that the World's Fair opened at the same space in Milan 1 month later.

Maison Et Objet

M&O is a major French trade show that is considered to be one of the most important design events in Europe that is held biannually in September and January. The 2015 September show promoted a theme of preciousness that re-examined the design community's concept of luxury from one of opulence and excess in mass and coloration to one of thin, directional lines with neutral pallets of slight blues, calm grays, excess yellowed orchres and bordeaux, terracottas and matte blacks.

Luxe Pack

Luxe Pack is the luxury packaging show that is held every year in Monaco, New York and Shanghai that features new innovations in packaging for the beauty, fragrance, luxury food and spirits markets. The Monaco venue is the original and largest show of the three. This is also a trade show where packaging design firms, and packaging manufacturers can showcase their new and existing product lines. Some of the trends for 2015 in New York were the multisensorial textures that were created by injection molding fabrics into plastics. Another trend was packaging that had dual functionality, changed color or temperature.

Innovation and Technology Shows

Technology Shows have become increasingly important since new technical inventions were previously lumped into other trade shows such as home furnishings shows, entertainment/music shows, automotive shows or fashion shows. Innovation conferences invite visionaries, futurists and inventors to discuss how the future will be impacted by new technologies and social movements.

CEATEC

CEATEC is the Japansese version of CES and is an acronym for Combined Exhibition for Advanced Technologies. Omron showcased their Kazoku Mesen surveillance cameras for children and pets that are styled as toy robots, mostly in gray with pink feet with accent colors of yellow, red, turquoise and orange. TDK introduced smart chopsticks that can sense the temperature and freshness of the food. These chopsticks were marketed by a company called Baidu, a Chinese search-engine giant. The chopsticks contain sensors that can detect hazardous food ingredients including recycled oils and unsanitary foods. http://blogs.wsj.com/chinarealtime/2014/09/03/is-your-food-safe-baidus-new-smart-chopsticks-can-tell/. Rohm Co. exhibited their flying paper crane. The origami structure operates from a smart phone is decorated in a traditional Japanese red paper with large yellow, green, white and blue flowers to attract attention.

Consumer Electronics Show

CES is the Consumer Electronics Show that is held annually in Las Vegas, Nevada. It started in New York City in 1967 as a spinoff of the Chicago Music Show which was responsible for featuring high-tech innovation related to music recording and playback. In 1998, the show relocated to Las Vegas and is one of the longest shows (18 days in total) in the world. In 2015, the first CES event was held in Shanghai. Some of the interesting products that were offered were Fitguard's mouthpiece that detects head trauma in athletes. The level of severity is indicate by a color change in the LED strip in the front of the mouthguard. No trauma emits a green light. Slight trauma emits a blue light and severe trauma emits a red light. Sensoria introduced a new pair of running socks that monitors a runner's gait that are offered in gray/black color. Another product that was featured was Rocketskates which are strap-on motorized skates that can exceed speeds of 10 mph and are available in red, blue, black and white.

Technology, Entertainment and Design

The TED (Technology, Entertainment and Design) conference started in New York City in 1984. The conference was started by architect Richard Wurman, as a one-off venue as the result of a trend that he identified as a convergence of ideas around technology, entertainment and design. Speakers include thought leaders in many different fields including music, science, mathematics, philosophy and philanthropy. In 2009, the venue relocated to Monterrey California due to an increase in attendance. The 2015 conference was held in Vancouver, Canada and attendance is only permitted to vetted members. Annual membership is approximately 6000 USD.

In 2012, TED started TEDx which are unofficial TED events that anyone can organize anywhere. There are a few restrictions including the conference must not make a profit but can charge an entrance fee to cover venue fees. Also the speakers cannot be paid. Speakers must also give up the copyrights to their presentations that belong to TED who is permitted to distribute the information using a Creative Commons license.

Creative Commons is a not-for profit founded by Lawrence Lessig in 2001 that establishes licenses for the free sharing of creative works so that others can build and expand upon it. The organization has released several copyright licenses open to the public which allow users to specify which rights they would like to relinquish and share.

Digital Life Design

Digital Life Design is a tech lifestyle conference (similar to TED) that started in Munich in 2005. It is an information exchange conference across various markets and interests including technology, creators, future visions, disruptors, and thinkers in the fields of innovation, science and culture. Previous speakers have included such luminaries as Mark Zuckerberg and Arianna Huffington. This is an invitation only annual event.

In 2010, DLD launched DLDWomen, declaring that the second decade of the new millennia is the decade of women and focuses on the gender debate and equality.

Globally, DLD has sponsored international conferences in Beijing, China, New York, Palo Alto, California and São Paulo, Brazil.

DLD also sponsored a summit festival in Israel in 2011 as part of the Tel Aviv Festival. The focus was to join Israel's digital media experts with an international business community.

99U

99U is a consulting business and conference that was founded in 2011 by Bēhance and is held annually in New York City, operating under the premise that 99% of what it takes to be successful resides inside of you (thus 99U). Bēhance is a network of websites that encourages self-promotion and is frequently used by students, artists and freelance designers to promote their work.

TechCrunch Disrupt

TechCrunch Disrupt (TCD) is an annual conference held in New York, San Francisco, Beijing and London that started in 2011 and is a venue for start-ups to present their ideas. The 2014 show featured some innovative products including MINK which is an inkjet printer that prints makeup, Mimi is a free hearing-aid device that runs off of an IOS app on an Iphone. The idea is to bring hearing enhancement to everyone who cannot afford a traditional hearing aid.

WSJD Live

The Wall Street Journal Digital live conference occurs every year in Laguna Beach California and is a tech conference that has big business feel. Some of the upcoming speakers include: Drew Huston, CEO of Dropbox; Michael J. Wolfe, CEO of Activate and Stewart Butterfield, Co-Founder and CEO of Slack.

Mobility Shows

Mobility shows are generally broken into the broad categories of Air, Boat, Sports Transportation and Automotive.

Automotive Shows

Automotive shows occur all over the globe and are pivotal to drive new car sales and automotive innovation. This section will examine global shows and international shows by regions. Currently, exterior automotive colors drive many industrial and consumer products for men's products and gender-neutral products. Interior automotive colors complement current ready-to-wear fashion trends.

Specialty Equipment Market Association Show

SEMA is the Specialty Equipment Market Association Show and is held annually in Las Vegas and is a global show. This is an after-market automotive parts show where vendors who create nonstandard automotive parts for cars right off the assembly lines show their products. As an example, pigment manufacturers might sell exotic or high-end pigments to traditional automotive paint manufacturers that will not be used for Ford or GM but might be used as an after-market paint to enhance the look or appeal of the mass-produced vehicle. Some of the products featured at the 2014 show were car wraps, 3D printed cars and after-market backup cameras.

North American Shows

Most larger US cities have an automotive show. Chicago hosts the largest show and features trucks and SUV's. The Los Angeles show features the most environmentally green cars and a wide variety luxury cars are an intrinsic part of the natural contrast for the California culture.

Detroit

The Detroit North American International Auto Show is the most important auto show in the world. It clearly favors the "Detroit Three" formerly referred to as the "Big Three." Some of the interesting trends in 2015 was Volvo's first-built Chinese vehicle to be sold in the United States: the Volvo S60 Inscription. The Inscription is available in seven exterior colors: black, two shades of white, red and three shades of gray. Another show stealer was the Ford GT350R Shelby Mustang. This high-performance vehicle with carbon-fiber wheels still favors form over function and is available in eight colors: red, yellow, blue, orange, black, white and two shades of gray. The 2016 Chevrolet Volt was redesigned to have a cleaner looking, more pared-down exterior. It is available in seven different exterior colors: red, two shades of white, black, two shades of gray and brown.

New York

Big cars are always on trend in the New York Auto show and Jeeps are also have a special place in the hearts of the New York audience. Most years, Jeep has a full-sized driving ramp outside the Javits center where the attendees can drive the Jeeps over rough terrain and experience the thrill of off-roading in an urban environment. The full-sized luxury brand from Cadillac; the CT-6 was a favorite at the New York show. The Cadillac Touring 6 is available in eight colors: two blacks, red, green, blue, white and two grays. The New York show also featured the down-market Jeep Renegade. This car featured several of Jeep's "Easter Egg" design details. Easter Eggs are little pictures of the signature jeep front grill with the round headlights and slotted grill. These designs appear in the tail lights of the Renegade as well as embossed in the speaker grille cover. The Renegade is available in nine colors: red, yellow, orange, black, white, two shades of blue, khaki green and gray. Toyota's RAV 4 featured a large plush pink Unicorn on its roof rack. This was a direct reference to Toyota's TV commercial that scripted a story of a family that carried the big toy home for the child in the back seat, indicating that the RAV4 is a family car that is capable of big tasks. The RAV4 is available in seven colors: red, two whites, black, two grays and brown.

European Auto Shows

There are several Automotive shows in Europe and the Paris and Frankfurt shows are probably the most influential. The Geneva automotive show started in 1905 and features many concept vehicles from many different countries.

Klimamobility

The Klimamobility show is held in Balzano Italy every year and features green and sustainable two, three and four-wheeled vehicles. Sponsors and exhibitors also include green technology companies such as battery manufacturers and fuel-cell innovators.

Frankfurt

The Frankfurt automotive show features cars from all over the globe but tends to favor German cars. Both BMW and Mercedes have their own halls that are lavishly decorated and frequently display racing vehicles and celebrities to draw in the crowds. Although the Frankfurt show occurs annually, the years alternate between commercial and passenger vehicles in conjunction with the Paris show. Although the 2015 show has not occurred yet, there is a tremendous amount of press about the 2016 Ferrari 488 Spider that will debut at the show. The car will be offered in the traditional Ferrari bright colors but also added a bright metallic blue to the lineup. BMW will be introducing their new 7 series which is reported to be almost 200 pounds lighter than its predecessor due to a metal to carbon-fiber replacement. The 7 series is available in 15 different exterior colors: six different black shades, six gray shades, two whites and blue.

Paris

Paris hosted the first automotive show in 1898 and since most consumers were skeptical of these horseless carriages, all autos were required to drive from Versailles to Paris to prove their mechanical feasibility prior to the show. The Paris Motor Show occurs every 2 years and the Frankfurt show exhibits passenger vehicles in the off years. The 2015 Bentley Mulsanne Speed was debuted at the Paris show and the manufacturer claims that it is as comfortable as an English manor house but with more assertive styling. It is available in 124 standard colors and combinations of colors as well as a custom color based on the customers wishes. Two new exterior colors are being offered, including "Spectre" a green-tinted black shade with a metallic gold effect and "Candy Red" which is translucent candy apple red shade. Piano black with carbon-fiber panels are also offered as a new color option for the interior.

Asian Auto Shows

The big three automotive shows that occur in Asia are Shanghai, Beijing and Tokyo.

Tokyo

The Tokyo Motor Show started in 1954 and was called the All Japan Motor Show. It still features the big three in Japan: Toyota, Honda and Mazda; however, there are many more manufacturers from China that are exhibiting as well as other manufacturers from around the world. In the early days, 90% of the vehicles exhibited were trucks and commercial vehicles. Many of the cars on display are concept vehicles. The Tokyo show featured Nissan's new Leaf Aero, an all-electric vehicle that is available in seven exterior colors: blue, red, black, two whites and two grays.

Shanghai and Beijing

The Shanghai International Automobile Industry Exhibition started in 1985 and now alternates yearly with the Beijing Automotive Show. The Bugatti Veyron 16.4 Grand Sport Vitesse Legend Black Bess debuted at the Beijing 2014 Auto Show. Only four of these vehicles were presented and all 12 that are planned for production have been sold out. The name Black Bess was inspired by an English Racehorse who was named after an old Bugatti 18 owned by French Aviator who flew a Morane Saulnier Type H airplane. There are hand-painted sketches of the airplane on the black interior leather of the vehicle and the exterior is completely black with 24-carat gold appointments.

Airshows

Airshows occur in many different parts of the world but the primary theme remains: the exhibition and technical capabilities of new, improved and vintage aircraft. A large draw at the airshows is pilots exhibiting their aeronautical skills. Another draw is the military displays of aircraft and prowess. Vast displays of innovation and luxury are always the highlights of these shows.

Dubai Airshow

The MENASA region (Middle East, Northern Africa, South Asia) is the aviation hub of the world and hosts the Dubai Airshow every 2 years. In 1986, the airshow started with a few dozen aircraft manufacturers to close to 200 manufacturers in 2015. The show has a strong military influence and the 2001 show started 8 weeks after the September 11th attacks. This show booked over 15 billion USD in record-breaking orders. In 2014, Mercedes-Benz Style and Lufthansa Technik provided 3D renderings of their latest VIP cabin interiors. The cabin was designed for about 12 passengers that exhibit and modern and elegant feel the featured color schemes were tone-on-tone neutrals, combined with an array of dark and light woodgrains. A minimal amount of chrome and brushed silvers were also in evidence. The design esthetic was a mélange of oriental and occidental style motifs. The cabin was outfitted in accessories of highly textured black and dark spice colors.

Paris Airshow

The Paris Airshow started in 1905 and is the world's oldest air exhibition. This venue occurs every odd year at the Le Bourget Airport north of Paris. Some of the trends that were featured at the 2015 show were the growing number of orders from Asian countries, further segmentation of economy, business and first classes split into five or six classes and expanded Internet access for passengers and the advances in cyber security needed for this access.

EAA AirVenture Oshkosh

The experimental aviation association's Air Venture Show in Oshkosh Wisconsin is the largest exhibition of experimental aircraft in the world. This event is held annually and started in 1953 and targeted builders and restorers of recreational aircraft. This is the premier event for "home-builds" where private aviators and innovators build aircraft from kits and scratch. A very common site at the 2015 show was the large parking lots of small planes with even smaller tents parked under their wings, housing the pilots and their families who flew in for the event. A full-scale exhibition of drone racing occurred in a large fenced in area of the show. One of the most interesting things to observe from a color perspective is the exterior paint schemes on the planes. There were elaborate design details on many of these planes but the most common colors were the primaries: red, yellow and blue followed by orange and silver.

Boat Shows

Monaco Yacht Show

The Monaco Yacht Show started in 1990 in France since Monaco is a premier destination for super yachts (over 75 feet in length). This show features ship builders, designers and brokers who come together to create and view the latest innovations in design. This is a good resource for color and textile trends for the luxury market.

Cannes Boat Show/Film Festival

The Cannes Boat Show and Yachting Festival is the premiere boating event in the Mediterranean. It is the largest in-water boat show in Europe and is also the home of the largest film festival in the world.

The boat show started because many film makers and guests would arrive to the film festival in a boat or would rent a boat while in Monaco for lodging or to stage parties or promotional events.

Monaco Grand Prix

Although the Monaco Grand Prix is a car racing event, yachts are available for charter for sailing, yachting, parties and watching the Grand Prix. Charter prices range from 5000 to 800,000 Euros for the week. This is also an interesting event to view the latest trends in luxury design and fashion because many of these boats hold fashion shows during the week.

Sports Transportation Shows

The Paris Motorcycle Show

La Salon De La Moto is a motorcycle show held in France every 2 years. This Cycle show started in 1901 and has had changes in venue and focus throughout the years. Initially, it was a bicycle and motorcycle show. Currently, it is considered the premiere event for motorcycles, scooters and quads in Europe. In addition to a venue that features motorbikes and scooters, it also exhibits motorcycle fashions and racing accessories that is interesting from a fashion trend perspective as well.

The London Bike Show

This is an annual exhibition for cyclists that features nonmotorized bicycles divided into the broad categories of road, off-road, family and urban. The show also features manufacturers of clothing, shoes and accessories for cyclists and their bikes.

Pacific Northwest Sportsmen's Show

This outdoor sports show is held every year in Portland Oregon in the United States. It is one of the largest sportsmen's shows in the world and was started in 1966. The show features fishing, hunting and camping equipment, vehicles, boats and clothing.

Further Reading

https://en.wikipedia.org/wiki/Consumer_Electronics_Show
http://www.huffingtonpost.com/2014/06/03/concussion-mouth-guard-fitguard_n_5399966.html
http://www.ceatec.com/en/award/award01_02.html
https://en.wikipedia.org/wiki/CEATEC
http://www.apartmenttherapy.com/whats-new-whats-next-5-hot-decor-trends-straight-from-the-runway-219685
https://en.wikipedia.org/wiki/Toy_Industry_Association#Trade_events
http://heimtextil.messefrankfurt.com/frankfurt/en/besucher/willkommen.html?nc
https://en.wikipedia.org/wiki/World_Wide_Web
https://www.pinterest.com/pin/377598749984979058/
http://www.planetdiecast.com/index.php?option=com_myblog&show=60-years-of-the-international-toyfair-in-nuremberghtml&Itemid=157
https://en.wikipedia.org/wiki/Behance
https://en.wikipedia.org/wiki/Digital_Life_Design
http://techcrunch.com/events/disrupt-ny-2014/
http://www.fierabolzano.it/klimamobility/de/
http://lesalondelamoto.com/en/visitors/about/history/
http://www.thesportshows.com/admin/wp-content/uploads/2011/05/01_2015_PNWS_Press_ReleaseFINAL.pdf
http://www.charterworld.com/index.html?sub=Luxury-Yacht-Events
http://www.thelondonbikeshow.co.uk/
http://www.prnewswire.com/news-releases/top-toy-trends-of-2015-announced-by-toy-industry-association-tia-the-official-voice-of-toy-fair-300036276.html
http://www.motortrend.com/auto-shows/sema/
http://www.motortrend.com/auto-shows/
http://www.edmunds.com/auto-shows/paris/
http://www.metropolismag.com/Point-of-View/June-2015/The-Best-of-NeoCon-2015/
http://www.spielwarenmesse.de/magazine/article-detail/trend-2016-design-to-play-toys-are-popular-with-parents-and-children/?L=1&cHash=3151474fce6e79e7314bd8b73be21a29
http://www.jsonline.com/business/pressrelease/national-press-releases/eaa-air-venture-oshkosh-2015-adds-more-highlights-to-the-worlds-greatest-aviation-celebration-311919951.html
http://www.cnn.com/2015/06/09/travel/le-bourget-2015-predictions/index.html paris airshow
https://en.wikipedia.org/wiki/Open_innovation
http://techcrunch.com/2014/05/06/mimi-launch/
https://en.wikipedia.org/wiki/TED_(conference)
http://www.ted.com/

29 Sporting Events

In recent years, sporting events have become an important directional in color trending fashions and accessories for men. Specifically, when pink disappeared from couture about 8 years ago, it was brought back into ready to wear and eventually back into couture through sports teams that were wearing pink accessories in support of breast cancer awareness. Initially, pink crept into menswear and boys wear (it never left girls wear) and then emerged back into ready to wear for women about 3 years later and found its way solidly back into couture about 2 years ago.

Summer and Winter Sports: Olympic Games

The Olympic games are the most publicized and viewed international sporting competitions in the world. At this point, almost every nation in the world participates and occurs every 4 years. The Olympics are divided into winter and summer games and alternate every 2 years. These events are highly politicized and fraught with controversy since cultures collide on and off the field. As expected, the fashions and uniforms that are made for the athletes are also the subject of intense media scrutiny and controversy. The opening ceremonies are similar to red carpet events with great press coverage and analysis. As an example in 2012, Ralph Lauren designed outfits for the US team, blue blazers and berets with white pants and skirts that were manufactured in China. Many Americans felt that the clothes should have been manufactured in the United States and felt that this was symbolic of outsourcing jobs crisis that many Americans were continuing to struggle.

American Football: The Super Bowl

The Super Bowl is the annual US football competition sponsored by the National Football League. Super Bowl I started in 1967 and has grown to become an enormous media event in the United States. It is

Color Trends and Selection for Product Design.
DOI: http://dx.doi.org/10.1016/B978-0-323-39395-9.00029-3

one of the most watched annual televised sporting events in the world and is only trumped by the European Cup (football/soccer) and the soccer game between the two teams from Spain: Madrid and Barcelona, the El Clasico. Many Americans feel that the commercials and half-time activities are just as important or even more important than the game itself. The day the game is played is commonly known as Super Bowl Sunday and is the second largest food consumption day after Thanksgiving day for Americans. After Super Bowl III where the New York Jets beat the unstoppable Baltimore Colts in 1969, Sports Illustrated published an article profiling the two quarterbacks: Joe Namath of the Jets and Johnny Unitas of the Colts. Unitas was the embodiment of his traditional masculine team that sports a blue horse-shoe on the side of the helmet whereas Namath was the embodiment of the modern American male for the NY Jets who was viewed as a jet-setter with his longer hair (Unitas had a crew cut) and wore full-length fur coats while frequenting posh clubs in New York City. The contrast between the old wild west of the Colts and the modern Jet-Age persona of the Jets was not lost on the readership of the Sports Illustrated Article and society at large.

Soccer/Football: World Cup

The FIFA Men's World Cup is the most watched televised sporting event in the world, where the final game is estimated to have a viewer-ship of almost 10% of the entire planet. The event is sponsored by the Federation of International Football Association and was founded in 1930. As with the Super Bowl, advertising sponsorship is extremely expensive but the impact of the brand owners is vast for sponsors such as Adidas, Coca-Cola, Hyundai, Visa and Budweiser beer. In recent years, World Cup Mascots have become increasingly important and are indicative of the spirit of the culture they are representing. An example would be the 1974 World Cup German Mascots: two boys, one with blonde hair, one with black hair dressed in black shorts and white shirts whose names were Tip and Tap. Another example is the American World Cup in 1994 a dog whose name is Striker (probably a Beagle) dressed in blue shorts and a red and white shirt. Some countries favored food mascots; in 1982 Spain's mascot was an orange (Narjito) and in 1986 Mexico's mascot was a red chile named Pique. There were also some more abstract mascots such as Italian mascot for the 1990 World Cup stick figure that was composed of red, white and green tiles named

Ciao and the 1992 Korea/Japan mascot named Spheriks which were three energetic monsters in purple, blue and orange.

Basketball: NBA

The National Basketball Association is an American organization that sponsors professional basketball teams and players. In terms of fashion and color influence, the largest area of influence is athletic shoes. Many of these high-end shoes focus on performance but are highly stylized and the styles are tailored to the preferences of the professional players that lend their names to the model of the shoes.

Sneakers emerged in the mid-19th century but the original, rubber-soled shoes were designed for the professional athlete. In 1917, Keds were introduced for everyday wear. In the 1930s, sneakers were associated with the pursuit of physical perfection, instituted by Adolf Hitler who encouraged his followers to excerise in service to the state. However, mass-production of the sneaker, lowered prices and availability, creating the first form of democratized form of footwear. In the 1950s, physical education became part of the curriculum for most school children and became part of their school wear. The gym rats and the physical fitness craze of the 1970s pushed the average consumer to achieve their "personal best." Those who could afford to join a gym or club demanded higher performance footwear for their exclusive activities, including racquetball, tennis and aerobics and jogging. The shoes were designed with colorful uppers and over-branded designs that indicated an elevated social status that was conveyed through conspicuous consumption. In the 1980s, Nike launched their first pair of Air Jordans and Adidas signed Run DMC who was the first musician to get a sneaker deal. This created a pivotal link between the athlete, the celebrity and the common consumer. These celebrity/athlete shoes elevated a trade shoe into fashion icon, mostly for men. In the 1980s, many women fashionistas were self-regarded as shoe-aholics but this new type of footwear became the domain of men known as sneaker heads. As the workplace became more casual, men traded in their wingtips for highly fashionable Air-Jordan's and later Le Bron James' or Kobe Bryant's. The shoes became identified by their celebrated athletes rather than their manufacturer. Basketball players were not the only people who lent their names to shoes: Spike Lee, Run DMC and many other rappers/music moguls were also a part of the scene. The first Air-Jordan's were red and black but later shoes exploded with colors and textures and were sported by even the most conservative men who not only wore them with athletic clothes but also denim jeans and suits. Sneakers are a pivotal part of modern men's fashion. It is not uncommon to see men on the runways of New York, Paris, Milan and Tokyo wearing couture with these celebrated athlete's shoes. Many of these shoe launches were made in limited production runs which elevated desire to obtain them. There were often news stories about teens being stabbed, mugged or murdered for their coveted shoes or about thugs lying in wait outside of stores, for the shoppers to come out with their purchases. In the 1980s, Sports Illustrated highlighted this phenomenon in an article, entitled "Your Sneakers or Your Life." The beauty and uniqueness of some of these shoes also elevated them to the status of art objects and are collected by sneaker heads who may have amassed several hundreds or thousand pairs in their private collections. Although many of these shoes were originally sold for a few hundred dollars, many have climbed in value into the thousands. In 2014, Rick Owens launched his "Vicious" line of clothing in shades of black, white, gray and taupe for men and women. No runway models were used but rather step dancers. Stepping is a type of dancing that is popular among black sororities and fraternities in the United States that involve clapping and stomping to make a powerful rhythm. Intense facial expressions convey strength and grit and meant to intimidate. The dancers wore Adidas running shoes designed by Rick Owens in matching shades of black white, taupe and gray. In 2015, the Brooklyn Museum launched a show called, "Sneakers" where they traced the history and cultural implications of the shoes.

Tennis: Wimbledon/US Open

Tennis is generally a warm-weather game so the idea of tennis whites was more of a practical choice to keep the athlete cool and hide perspiration stains rather than a fashion trend. Early men's tennis clothes included sweaters, vests and long pants. In the 1930s, men started wearing white shorts rather than flannel pants. When women joined the game, they wore bustles, corsets and ties, all in white. As society's fashions have become more spare and fitted, the tennis outfits have also become shorter and more lean. Men opted for lighter, more technical fabrics such as Lycra and nylon shirts and shorts that hug their bodies and celebrate their athletic form. Women started wearing shorts and short skirts in the 1940s and 1950s. A technical problem was encountered when tennis became widely televised in the 1970s and announcers complained that they could not tell the players apart since they were all wearing white so the USTA allowed players to wear pastels in the US open. In 1981, John McEnroe won the Wimbledon singles tournament while sporting a patriotic ensemble of a white shirt with blue stripes, a red head band and white shorts. The tournament was held on the fourth of July. In the 1980s, a female player wore a white unitard to Wimbledon that her opponent found distracting. She was ordered to change her outfit for the next day. Chrissie Evert's diamond bracelet fell off her wrist, stopping the game at the US open, and started the tennis bracelet trend with Tiffany jewelers. There is a certain "je nais se quois" about the powerful Williams sisters that is undeniable in the fashion world. Venus and Serena Williams quickly became fashion icons for the tennis world starting with their beaded hair. The beads frequently fell out of their braids during matches and the sisters could be seen swatting them off the court with their tennis racquets. In 2002, Serena appeared at the US Open in a svelte black glossy catsuit. In 2007, Maria Sharapova wore a red Nike ensemble that was adorned with crystals outlining the NYC skyline to the US Open.

Further Reading

http://espn.go.com/olympics/summer/2012/story/_/id/8160403/2012-olympics-china-made-team-usa-uniforms-ignite-anger
https://en.wikipedia.org/wiki/El_Cl%C3%A1sico el classico
http://nymag.com/fashion/08/fall/49260/ tennis uniforms
https://www.washingtonpost.com/news/arts-and-entertainment/wp/2015/07/10/how-a-sexist-sneaker-culture-turned-men-into-fashion-addicts/
http://library.la84.org/SportsLibrary/JSH/JSH1988/JSH1501/jsh1501d.pdf
https://en.wikipedia.org/wiki/New_York_Jets#Logos_and_uniforms
http://olympicfashion.tumblr.com/
https://en.wikipedia.org/wiki/Super_Bowl

30 The Media

The media has an enormous influence on the fashion world as well as the fashionable consumer. Although current color and style trends are harder to see than those from TV shows, popular magazines or movies, if one pages back a few decades, the trends are much easier to spot. This is due, in part, because the media that survives is more compressed and the trends are more distilled and iconic in retrospect.

TV Shows and Music Videos

There are many TV shows that intentionally highlight fashion trends and celebrity. One of the earliest cable shows was produced by the Entertainment Network called, "Fashion Police." This is a panel of three or four fashionistas. The original panel included Joan Rivers, who was a comedian and an outspoken proponent of couture as well as Kelly Osborne, Ozzy Osborne's daughter who is fond of alternative fashions, hairstyles and colors. This group regularly reviews the fashions of celebrities at red-carpet events as well as featured photos in tabloids. They are called the fashion police since they have very rigid rules of what people should and should not wear. These types of pronouncements have an impact on fashion designers, stylists, celebrities and society at large.

Many reality TV shows such as the Housewives series on Bravo and some of the Real World series on MTV do not tell the audience about fashion but rather highlight what the desirable segment is currently wearing.

The Learning Channel aired a show called, "What Not to Wear" which targets a fashioned-challenged unfortunate who is shamed by his/her colleagues and family to appear on the show and receive a makeover. Many fashion and social rules are expressed and implied about matching colors and textures of clothing, good and bad hairstyles and haircolors, makeup choices and appropriate shoe selections for different activities. Most of these are culturally biased towards the Western sensibility and should appear outlandish in a few decades.

Color Trends and Selection for Product Design.
DOI: http://dx.doi.org/10.1016/B978-0-323-39395-9.00030-X

Many TV series choose the colors of their actors costumes and settings in deliberate and thematic method. As an example, the recently popular, "Breaking Bad" series on Netflix consciously chooses colors and color names of characters to provide depth and establish relationships between the characters. The protagonist, Walter White, has an alter-ego, Heisinberg (a drug dealer) who is frequently dressed in green clothing or is illuminated in a greenish hue. This color was deliberately chosen since it is the color that represents money, greed and growth. Walter's wife frequently wears blue and Walter's drug partner frequently wears yellow, the combination of the two equals green. Walter's baby that was born during the series was named Holly White which is the common name for an evergreen shrub. Walter is frequently dressed in blue and illuminated in a bluish hue. This color was chosen because it signifies sadness, power and loyalty. The high-end meth product that Walter produces also has a signature blue color known as "Blue Sky"; however, typical meth products of lower quality are usually a yellowish crystal. Many of Walter's competitors, enemies and distributors wear yellow.

Sex in the City was a television show that aired on HBO in the early 1998–2004 that was a romantic sitcom about four middle-aged women set in New York City. The protagonist, Carrie Bradshaw, was writer who wrote a sex column for a New York paper and was a recognized member of the New York Glitterati as she frequently attended social events and club openings where she met men and subsequently had relations with them. Fashion was a frequent topic on the show as well as sex and relationships. The show created a romantic notion of New York's fashion and dating scene that was not quite as visible to the outside world prior to the release of the show.

Television Commercials

TV commercials were extremely important in the 1960s and 1970s but with the advent of nonnetwork, cable channels that show fewer or no commercials, they have lessened in importance. The exception is the commercials during the Super Bowl and other major sporting events including the Olympics. One genre of programming that continues to be influential is children's programming. These commercials are the subject of great criticism since their undue influence on impressionable young minds is sometimes considered manipulative, espousing the great taste of processed junk foods as well as an endless array of toys and gadgets that never seem to function as well as they did in the TV commercials.

Print and Media Campaigns

There are many basic rules of color selection when it comes to print media. Many of these are practical and relate to the relative ease of viewing and message conveyance. As an example, using light colors on a white or light background makes print hard to read. A red background with blue lettering is also difficult to read since the colors tend to bleed together. Also using a textured or busy pattern as a background for print can be visually distracting. This can also be true of an over-use of bright colors or too many colors in a print campaign can be confusing for the reader. Fluorescent or neon colors on a dark or black background can also be visually irritating. It is best to use less than three colors if possible. In general, strong colors like black or intense colors like bright reds, yellows or greens should be avoided for background colors since they can cause visual fatigue.

Magazines: Advertisements and Specialty Editions

Vogue Fashion

Vogue is a fashion and lifestyle magazine owned by the Conde Nast publishing empire and was founded in the United States in the late 1800s. The magazine was initially targeted to the New York elite and featured articles about lifestyle, fashion, sports and politics for both men and women and was initially published as a newspaper. When Conde Nast bought the newspaper a few years later, he changed the format from a weekly newspaper to a bi-weekly magazine. Currently Vogue is published as separate editions in 25 different countries and regions. The first two editions that were added were United Kingdom and France in 1920. The most recent editions that were added are Thailand and the Ukraine in 2013. Germany was added in 1924 but ceased publication from 1944 to 1979. There is also a Teen Vogue, several men's Vogues in different countries as well as a bridal Vogue, a bridal Vogue, a baby Vogue and home Vogue among many others.

There is a political undercurrent to these publications both within their publication as well as a mirror to the politics occurring globally. The American, flagship publication has been a center of controversy and scandal concerning the editor-in-chief. Each of these women is not only known for their fashion and business acumen but also are celebrated personalities in their own right. The last three were probably the most celebrated: Diana Vreeland, Grace Mirabella and Anna Wintour. Diana Vreeland was the Vogue editor-in-chief from 1963 to 1971. Her apartment in New York was painted in a vibrant red color that she decorated like a garden from hell. She also painted the walls of her Vogue office in the same red color, outfitted with a leopard skin rug. She was a fashion consultant to Jackie Kennedy for the inauguration and later became the pivotal director of the Metropolitan Museum's Costume Institute in 1972 after she was fired from Vogue due, in part, to her lavish photo shoots and parties. Vreeland was quoted with such statements as "pink is the navy blue of India." In contrast, Grace Mirabella, who became editor-in-chief in 1971, was not a socialite but rather a disciplined retailer who tripled Vogue's circulation during her tenure. She painted over Vreeland's red office walls and redecorated in tones of beige. Mirabella was replaced by the much younger socialite Anna Wintour 17 years later. Anna Wintour took Vogue by a storm, firing many of the editors, changing photo shoots from head shots to half and full body shots and restoring Vogue to its former majesty and flamboyance of its earlier history. Wintour cut a formidable figure with her couture style of dressing along with her signature bob haircut and large sunglasses. In 2014, Forbes magazine named Wintour as the 28th most powerful living woman in the world. This is an interesting placement since Angela Merkel, Hillary Clinton and Melinda Gates were named #1, 2 and 3, respectively. Queen Elizabeth II was #41, Ellen DeGeneres was named #50 and followed at #100 is Lee Bo-Jin, heir apparent of the Samsung Group in Korea.

WSJ

WSJ is the luxury fashion and lifestyle magazine of the Wall Street Journal that was launched as a quarterly in 2008 and became a monthly in 2014. In 2013, Ad Week awarded the WSJ "Hottest Lifestyle Magazine of the Year." In 2012, they initiated their Innovator Awards that honors the top creative performers in the fields of design, fashion, art and technology. Two of the recent winners were Pat McGrath for Fashion and Thomas Woltz for design.

New York Times Style Magazine

The New York Times Style Magazine is a publication that is inserted with the Sunday New York Times in the United States that is published

13 times per year that was launched in 2004. It covers topics ranging from art and design to fashion and lifestyle, to food and culture. It is also distributed with the International New York Times weekly edition since 2007. An example of a recent article is a piece about Ai Weiwei's new wearable art collection. Ai Weiwei is a dissident artist from China who has been imprisoned for his outspoken beliefs of injustice an inequality in China. His new jewelry line is a collection of gold bracelets that look like small pieces of twisted rebar. Rebar is a reinforced metal bar that is frequently added to concrete structures to reinforce the strength of the building or object. The jewelry line was influenced by the devasting earthquake that occurred in China where almost 100,000 people were killed. Ai Weiwei visited a school where many children were trapped under the rubble that was aptly dubbed a "tofu-dreg school." This term comes from the idea that the left over pieces of tofu after processing are cobbled together to make an inferior product. This is a good example of a piece of journalism that lies within the intersection between art, design and politics and how they influence each other. Ai Weiwei's 24 karat gold rings, bracelets and necklaces were for sale for a few months in 2015 at the Elisabetta Cipriani Gallery in London, prices ranging from 35,000 to 70,000 USD.

New York Times Style also has an Internet site with articles, videos with commentary and also an Instagram page called New York Times Fashion that has over 1 million followers. There are images currently on Instagram about denim and how denim pant sales have been sliding for the past few years but are now picking back up due to the purchasing power of the millennials.

Sports Illustrated Swim Suit Edition

The Sports Illustrated Magazine is a weekly publication by Time, Inc. When the magazine started in the United States in the 1930s, the emphasis was on the sportsman and the focus was on activities such as yachting, golfing and tennis. Over the years, the emphasis shifted from the sports of sportsmen to the sports of spectators. In order to accurately cover the schedule of spectator/professional sports, the magazine needed weekly rather than monthly or bi-monthly publication. The winter months are notoriously slow for spectator sports so the swimsuit issue was added to help filling this gap. This issue features female supermodels on the cover wearing the latest swimsuit fashions and hairstyles. Since 80% of the readership is male, this is a unique venue to introduce style and fashion trends to men. There has been much self-generated controversy about this edition being defined as thinly veiled pornography. Although, the negative feedback from the readership has been minimal, SI has been accused of over-inflating this feedback to fuel sales of their magazine and increase readership. SI has successfully established the link between sex and sports through the use of fashion trends.

The Internet

It would be presumptuous to assume that some readers need a definition of the Internet as an electronic network of networks since its presence and influence is insidious in almost every facet of the average consumer's experience. The Internet was developed in the 1960s by the US government to develop a communication network to share information for the government and the military. The World Wide Web is the primary tool that consumers use to access the Internet. Tim Berners-Lee developed the first web browser in the early 1990s when he was an employee of CERN (home of the Hadron Collider) in Switzerland. There are numerous trend spotting sites that designers, consumers and product manufacturers visit to understand the latest information and products that are available for a specific market of purpose.

Kickstarter

Kickstarter is a crowdfunding site that enables startups and individuals obtain funding for projects that might not be funded by traditional means such as bank loans or grants. Kickstarter was launched in 2009 soon after the great recession of 2008. Their mission is to make creative projects commercially viable through online funding. The backers of these projects receive gifts ranging from branded t-shirts to early or pre-launched products. What started as individual contributors vetted by a panel at Kickstarter, has evolved into groups of curated projects by many prestigious organizations such as TED (technology entertainment design), Art Basel and Rhode Island School of Design (RISD). Kickstarter is also a resource for new product development as well as trend tracking for new social movements, materials, technologies and colors.

Instagram

Instagram (instant camera/telegram) is an app that allows consumers to post photos, videos and commentaries about whatever interests them onto social media sites such as Facebook and Twitter using hashtags as a meta filtering device to link common themes, trends or link like-minded users. As an example, Instamoms are mothers that use social media to post pictures and videos of their children for social promotion and/or to generate revenue. Social media is creating a new source of Mom managers. In the past, the traditional stage mother was someone like Kris Jenner who took her girls into modeling agencies and helped to negotiate their contracts and salaries. The Instamoms are by-passing the modeling agencies and are taking their children directly to the public, sharing their personal trends and influencing millions of viewers and followers directly.

Blogspot

Blogspot is a subdomain of Google where users can log on and write about any subject (almost) of their choosing that was launched in 1993 by Pyra labs and was purchased by Google in 2003. Users can also post links, photos and videos. Many users write blogs to promote a cause or product, share their daily activities or make money through reviews or contacts. Since it is now part of Google+, there are many opportunities to link Blogspot (Blogger) to other applications such as Amazon, You Tube and Microsoft Word. This site is blocked in at least 15 different countries including China, Russia and many Middle Eastern countries.

Twitter

Twitter is often regarded as the SMS (short messaging system) of the Internet world and is a global platform for real-time self expression and connection. It started in San Francisco in 2006 as a communication tool to keep family and friends updated. It allows users to send and read 140 (maximum) character updates or tweets. From a trend perspective these sound bites can be quite interesting because many perspectives on one event or product can be obtained rapidly but with no depth or context so trend information can be easily skewed or misinterpreted. It should be used more as a point of validation rather than a source for trend information.

Tumblr

Tumblr is a microblogging social network service that was founded in 2007. This site differs from Blogspot since the entries or posts are quite shorter but still longer than Twitter. Tumblr is owned by Yahoo.

Pinterest

Pinterest is a website where users share photos by pinning them to their boards that was launched in 2010. The individual pictures or links to pictures are referred to as pins and the pins are collected or pinned onto boards where they can be viewed by other members. These boards are generally categorized by theme and some of the more popular themes are categories like weddings or specific holidays, recipes, vacation spots, etc. Pinterest members are mostly women. The latest global demographics suggest that women comprise somewhere between 70% and 90% of the viewing audience. However in Great Britain, more men are users than women. Many retailers use Pinterest for trend research and to sell products. As an example, there is a company called Trend Council which is a fashion trend consulting firm that has pinned their spring and summer colors for 2017 on the color trend board. They feature pink, orange, yellow, aqua blue and green color themes. Each theme contains five swatches of these color families.

TED Talks

As mentioned previously TED conferences occur as live venues all over the world. Attendance is extremely limited and quite costly so the TED group developed a website called TED.com that posts over 2000 of their speakers from the conference so that millions of viewers can watch the proceedings for free. In 2009, they have also initiated their open translation project (OTP) that makes their talks accessible to the billions of non-English speaking people. This program makes use of crowd-based subtitling for the talks to be translated into almost any language. Thus far, talks have been translated into over 100 languages by volunteers.

Further Reading

http://breakingbad.wikia.com/wiki/Color
http://www.slate.com/blogs/the_slatest/2015/10/03/syrian_refugees_paris_fashion_shows_and_the_supermoon_in_this_week_s_best.html

http://nymag.com/nymetro/shopping/fashion/features/n_7930/index1.html vreeland

http://www.nytimes.com/2015/11/12/t-magazine/ai-weiwei-first-jewelry-collection.html?_r=0

http://www.nytimes.com/1995/09/24/books/grace-under-pressure.html?pagewanted=all&src=pm

http://www.nytimes.com/2015/11/15/fashion/meet-the-instamom-a-stage-mother-for-social-media.html?&moduleDetail=section-news-2&action=click&contentCollection=Fashion%20%26%20Style&region=Footer&module=MoreInSection&version=WhatsNext&contentID=WhatsNext&pgtype=article

https://instagram.com/p/9uNBiXLtBc/?taken-by=nytimesfashion

http://www.slate.com/articles/news_and_politics/the_middlebrow/2005/02/the_sports_illustrated_swimsuit_issue.html

https://en.wikipedia.org/wiki/Sex_and_the_City

http://www.wsj.com/public/page/magazine-index.html?mg=blogs-wsj

http://wsjdconference.wsj.com/

Berners-Lee, T., Weaving the Web. Harper San Francisco; ISBN 0062515861 (paperback), September 22, 1999.

31 The Hot New Color

Frequently, down market suppliers will inquire about what is the "Hot New Color." This is usually a very short-sighted question and asked by those who have a limited understanding of color direction and trends but need color direction to offer to their upmarket customers. The idea of a Hot New Color implies that one color will be on trend or popular with every application for every industry and convertors should offer this color to all of their customers regardless of their product line. This can lead the customer to infer that the supplier is limited in their ability to supply innovative products or provide suggestions about how to refresh or enhance their existing product lines.

Color of the Year

Recently, several color manufacturers and color measurement companies have been releasing their, "Color of the Year." This is a very myopic approach to color direction since it seldom segments markets, color compatibilities and trend catalysts. This is another common mistake similar to the hot new color phenomenon; however, there is some validity to certain shades of colors that have been more popular that year but it will not work across different industries.

Color With Context

The type of color trending that is published by groups such as the Color Marketing Group (CMG) is far more relevant than color of the year since they bring together color professionals from many different industries to discuss emerging and future color trends. This can also be insulated and somewhat self-serving since many of these professionals work for color companies or companies that make colored products that they want to promote. However, there is a series of checks and balances in these types of workshops where some industries are far ahead (or behind) of the trends of the other industries and so these types of

Color Trends and Selection for Product Design.
DOI: http://dx.doi.org/10.1016/B978-0-323-39395-9.00031-1

discussions may help to determine what is a pasttrend or an emerging trend. All of these discussions engender a great deal of context by the members who must defend their colors before their peers prior to the voting process so when these color professionals return to their studios, they are armed with a pallet and a justification for the choice of those colors that enable them to create pallets for their own work that is cohesive and contextual.

Further Reading

http://www.cnn.com/2014/12/05/living/pantone-color-of-the-year-2015-marsala/index.html
http://www.printmag.com/article/flaming-pink-the-hot-new-color/
http://www.colormarketing.org/

32 Color Trends and How They Influence Color Selection

Each designer, artist, marketer or engineer must ultimately land on a colorway pallet or selection for their products. Sometimes this has nothing to do with trends. Frequently, the pallet is limited to brand or trade colors that is part of their product line. In some cases, certain colors will be off-limits due to their stability in or on a certain type of product. Other times the project is a color refresh where the colors might be the same hue but are changing in chroma. In other cases, an engineer might be asked to find a lower cost alternative to an existing product to support the growth of a mature product. This is frequently the case when a product contains more expensive pigments such as pearlescent or color-shifting products that draw initial attraction to the product but wanes in importance once the efficacy of the product is established and the consumer re-purchases the product. A good example of this is haircare products that frequently launch with expensive pearlized packaging to convey the high-end quality of the product to the consumer and once the product has matured, the manufacturer, redesigns the color without the pearl and creates a more commoditized package to enhance profit as sales volumes decrease. It has been argued that as more consumers purchase products from their computers rather than shopping in brick and mortar stores, the colors used in products and their packaging becomes less important.

Further Reading

http://brandcolors.net/
http://www.hgtv.com/design/decorating/color/color-trends-whats-new-whats-next

Color Trends and Selection for Product Design.
DOI: http://dx.doi.org/10.1016/B978-0-323-39395-9.00032-3

33 Color Language

The language of color is as specific as any other language but frequently can transcend barriers between people who speak different languages. Also the words that an engineer might use to describe color are usually quite rational or technical and are related to color measurement or optical properties of colors whereas the words that designers might use are much more sensorial or emotional and relate to how a color feels or appears or how it relates to its surroundings. As an example, an engineer might use words like chroma, saturation, hue, opacity, etc., whereas designers might use words like warm, clean, invigorating, smooth, etc.

Brand Colors

Brand colors are most simply described as colors that help to define or reinforce the brand. Sometimes these single colors or combinations of colors are owned by the brandowner and are legally protected from other brands using these colors. In the United States, there are three broad types of copy protection: trademark, registry and copyright.

Trademark (™) is an unregistered trademark for products, symbols, names, designs and images so they will not be used by competitors to gain an unfair advantage. A typical trademark does not protect the color or color scheme of the design but rather just the design itself. A color trademark is somewhat different. In this instance, the color is the brand of the product. An example of this would be the yellow color of Post-its used by 3M. Although 3M produces Post-its in many different colors, they promote their brand with the original yellow shade and have protected it with a trademark. John Deere began trademarking their logos in 1876. The logo was printed or painted in black over a white surface and evolved into a simpler and stronger logo over the next 100 years. In 2000, they trademarked a new logo simple deer green coloration, contrasted with yellow trim. The US government permitted them to trademark the logo but was unable to protect the green color because it was determined to be a color in use or a functional color. Since green is

Color Trends and Selection for Product Design.
DOI: http://dx.doi.org/10.1016/B978-0-323-39395-9.00033-5

so closely associated with agriculture and growing plants, it was considered to be a functional color and that John Deere would be given an unfair advantage if their competitors were unable to use the color green to represent their products. John Deere does have their own shades of yellow and green paint that they own that cannot be used by their competitors. Trademarks can also be given to other unconventional items other than color such as scent or a sound (ringtone). Any items that are considered offensive cannot be trademarked.

A registry (®) is a trademark that has been registered with the USPTO (United States Patent and Trademark Office). This registration is only granted to images, designs or products that are considered unique. As an example, Coca Cola red is a trademarked color but only for similar products. The same is true for Tiffany blue for jewelry packaging and associated products. Owen-Corning pink is also registered but only for fiberglass. Christian Louboutin filed suit against Yves Saint Laurent to prevent YSL from adding a red outsole to the bottom of their stilettos. YSL was permitted to use the red outsole for red shoes but not shoes of any other color.

Copyrights (©) are somewhat different since they are commonly used for publications, creative artwork and images. This allows the owner of the copyright to copy, reproduce, distribute and sell these images and prevents others from doing the same without the permission of that copyright owner. Currently copyrights can protect images and artwork but it does not protect the colors used in those images or artwork.

Creative Commons were discussed briefly under the TED Conference section and are one of many public copyright licenses that have a variety of restrictions and allowances depending on the type of license used.

Further Reading

http://www.differencebetween.net/business/difference-between-tm-and-registered-trademark/

http://www.newsweek.com/footwear-sole-sexiness-140123

http://freakonomics.com/2011/08/12/can-you-trademark-a-color/

https://www.deere.com/en_US/corporate/our_company/about_us/history/trademarks/trademarks.page

34 Color Descriptions

Color descriptions can serve as short cuts for an actual image of a color. This book will have no color plates or renderings because it relies solely on the use of color descriptions and names. This was done intentionally to eliminate color variations in inks as well as fading or aging of colors over time. As an example, when something is described as having a pinkish hue, that is quite easy to visualize but if a color plate is printed with a pinkish hue, that could vary significantly in hue and chroma from one printing to the next as well as in digital renderings or if the ink colors fade or yellow over time. Frequently, designers and engineers will talk about a color shifting to a different hue. As an example, a designer that is trying to achieve a turquoise color in a finished product might find that the turquoise shifts too far to the yellow or that the color needs to be a little less blue. Perhaps the color is too saturated and needs to be more of whitened pastel shade of turquoise. Perhaps the color is too washed out and needs to be more intense or have a higher chroma. It always best to have these color corrections or discussions with the stakeholders in the same room, looking at the color using the same light source, at the same viewing angle with the same background but this is not usually possible so many times color language is used to highlight differences in colors from their targets or to describe a color within the context of another color.

Color Names

Color names can be a slippery slope when it comes to accurately describing a color. A good example is the name Chocolate to describe a brown shade. This can be terribly misleading because chocolate can be almost any color from white to orange to brown to black. Some color names are better than others. Cocoa might be a better descriptor for brown than chocolate. Violet or eggplant might be better color names than purple. However, most color designers name their colors because it

Color Trends and Selection for Product Design.
DOI: http://dx.doi.org/10.1016/B978-0-323-39395-9.00034-7

is easier to catalog and track the colors that they have developed or sold to customers. There are some color descriptors that are quite accurate. A good example is Robins Egg Blue where the reader can generally imagine the color quite accurately as opposed to a color name like bone or bark.

35 International Appeal

As mentioned previously, there are different cultural and demographic biases for certain colors. However, there are some universal colors that will sell in most regions. Neutral colors such as tan, off-white, gray and muted shades of blue will work in almost any region. Golds, silvers and pearls are considered universal high-end or luxury colors. Blacks, metallic grays, blues and whites are generally regarded as high-tech colors.

Further Reading

http://www.thechicfashionista.com/neutrals-universal-colors.html
http://www.onextrapixel.com/2010/01/22/anatomy-of-colors-in-web-design-blue-and-the-cool-look/

Color Trends and Selection for Product Design.
DOI: http://dx.doi.org/10.1016/B978-0-323-39395-9.00035-9

SECTION III

Color Innovations and Emerging Technologies in Color Development

New Technologies for color development and color applications are essential to understand the esthetic and technical pallets of applications, colorants and color measurement before deciding on production methods and colorants for new products.

36 Principles of Color Science

It is important for designers, specifiers and engineers to understand how their colleagues, competitors and other people in different industries look at color and understand the basic definitions of colors and color science.

Colorants

Colorants are simply defined as component that adds color to another material. Generally colorants can be broken down into dyes and pigments. For the sake of simplicity, we will look at dyes as one category and split pigments into organic pigments, inorganic pigments, metallic pigments, pearlescent pigments and metallescent pigments.

Dyes

The primary difference between pigments and dyes is their ability to dissolve within the matrix of a product. Generally speaking, dyes when introduced in a molten plastic or liquid paint form a solution. On the other hand, pigments form a suspension when introduced into a paint or plastic matrix. Since dyes lose their crystalline structure when they are introduced into a matrix, care must be exercised to insure that the matrix can hold the dye and not allow it to migrate to the surface of the product. A good rule of thumb is to use pigments for olefin-based (crystalline) polymers and dyes or a combination of dyes and or pigments for nonolefin (amorphous) polymers. Dyes have a higher tinting strength than pigments and are much less stable than pigments. Natural dyes have been used since the beginning of time to color textiles, cosmetics and paints. Many natural dyes were derived from plants and animals. As mentioned previously, cochineal insects are native to Latin America and are commonly referred to as cactus louse. The female of the species creates a red liquid that is extracted from her carapace to create a red dye. This red color can shift from a light pink to a brown all the way to a deep blue color depending on

Color Trends and Selection for Product Design.
DOI: http://dx.doi.org/10.1016/B978-0-323-39395-9.00036-0

the pH of the solution. A similar insect in India called Kermes also yields a red dye that also shifts color upon exposure to an acidic or basic media. Lac Dye is another red that is excreted from the shellac resin of the Lac insect in Southeast Asia. Red dyes are also obtained from Madder. The plant root from the *Rubia cordifolia* is the source for the Indian Madder Red dye Tumeric and saffron yield yellow and orange dyes and the indigo plant yields a strong blue colored dye. Indigo dye is made by a vat process where the indigo plant is fermented and then treated with urine. Fibers are then dipped into the colorless solution and allowed to dry in the sun to develop the blue color and to ensure color fastness. Tyrian purple is derived from a predatory sea snail from the Mediterranean Sea. Although synthetic dyes are relatively new, they have been in use in various forms for several hundred years.

The first synthetic dye on record was created a Dutch chemist, Cornelius Drebbel, in 1630 by combining natural cochineal and stabilizing it with tin. He also invented the first navigable submarine vessel in 1620. In 1774, Prussian Blue was developed by combining the prussite of potash with the iron salt of copper. Much controversy still exists about who was the inventor of this dye. Synthetic indigo dye was by Johann Von Baeyer in 1865 and in 1888 BASF created the first viable manufacturing process for synthetic indigo that is still currently in use. By 1913, BASF's synthetic indigo had almost completely replaced the need for natural indigo. The largest market for the indigo was to dye the textile denim that was first used for sails for ships and later for sailor's pants. The sailors of Dhunga India were responsible for the new name dungarees whereas the term jeans came from Italian laborers in Genoa who preferred the denim pants over other textiles due to its durability. The use and invention of synthetic dyes expanded rapidly as synthetic textiles such as rayon and nylon were developed in the late 1800s into the 1900s.

Organic Pigments

Organic pigments were first developed about 300 years ago. The simplest definition is that these pigments consist of conjugated ring structures of Carbon, Nitrogen, Hydrogen, Oxygen and Sulfur atoms. Like most dyes, organic pigments are quite intense in color but they are much more stable due to their attachment to an inorganic salt. They also have a high tint strength and yield clean color shades.

Inorganic Pigments

An inorganic pigment is a synthetic or natural metal salt or mixed metal oxide that can be used to color plastics, textiles, paints or inks. Inorganic pigments are generally much duller in appearance when compared to their organic counterparts but are much more heat and light stable. Titanium Dioxide or Titanium White was discovered in the early 1800s but became a mass-produced artist's pigment in the 1920s. This pigment, along with the refinement of acrylic paints in the 1950s, shaped the color pallet of the modern art movement.

Special Effect Colorants

Special Effect Colorants includes a whole host of pigments and dyes that are both organic and inorganic in structure that impart more than just chroma and hue to a product but also contribute other visual properties such as reflectivity, luster, shimmer, sparkle, pop, fluorescence, phosphorescence, etc.

Metallic Pigments

Metallic Pigments are, most simply, ground up pieces of metal that are suspended into plastics, paints and inks. Some of the most common metals used are aluminum, iron, copper and gold. These pigments are quite opaque and provide good hiding and high reflectivity when used properly.

Pearlescent and Luster Pigments

Pearlescent pigments were discussed briefly in a prior section. Pearlescent pigments and special effect pigments can be subdivided into further categories: white pearlescent pigments, interference pigments and metallescent pigments.

The first pearlescent pigments were created from pulverized fish scales. Natural pearlescence was commercially available in the early 1900s when fish scales were removed from Herring and the guanine and hypoxanthine crystals were extracted from the scales and sold as a dispersion for cosmetics and high-end automotive paints. Although this created a beautiful finished product, the process was very labor-intensive and the gathering of the fish scales could only occur a few months in the year. In the 1950s, a new product, synthetic product called lead carbonate became commercially available and was widely used well into the 1970s but was phased out due to the toxicity of the lead component. This pigment is still available for use in Asia today and is used to manufacture garment buttons and eyeglasses that are still imported into the United States. In the 1970s, two other products were developed to replace lead carbonate pigments: bismuth oxychloride and titanium coated micas. Bismuth oxylchloride pigments also yield a very strong luster effect but tend to photodarken when exposed to sunlight. Many manufacturers have added a UV stabilizer to their formulations but has made the product cost-prohibitive. Bismuth products are still commercially available but are used primarily for cosmetic applications, mostly for nail enamels. Bismuth product is also still used to make bowling balls. Titanium coated micas are platelets of mineral, mica (potassium, aluminum silicate) that are coated with a thin layer of titanium oxide. These products are available in a large range of particle sizes and are quite photostable when compared to bismuth or natural pearl pigments. The one drawback is that they are extremely shear sensitive due to their platey nature. Natural mica is mined from the earth in very large sheets that are quite friable when stressed. This is generally not a problem for paints and inks but when extruded or molded into plastics, they can easily delaminate, stripping the titanium oxide coating off the surface and losing all of its luster. Care must be exercised to introduce these pigments downstream in the plastics processing machinery or use low shear equipment that does not compromise the integrity of the coating. Since these natural micas are comprised of flakey structure that can split into thinner layers with excess shear, the interference or pearl color can be reduced or completely grayed out of the pigments become overly fractionated. Another common issue with mica-based pigments is the occurrence flow, weld and knit lines in molded plastics. Although mica pigments do not cause flow lines (flow lines are caused by the configuration of the mold and how the plastic flows into that mold), they tend accentuate the molding flaws that would not be evident in the finished pieces that contain other types of colorants. It is beyond the scope of this book to examine how to reduce or eliminate these molding imperfection but a general rule of thumb is to use larger, more translucent pigments with a less-platey morphology.

Interference or multicolor pearlescent pigments are mica-based pigments that contain a thicker coating of titanium on the surface. The thinnest coating produces a white pearl, followed by a blue and yellow/gold pearl with a green pearl having the thickest coating of titanium.

In the 1990s and the early 21st century, other substrates were employed to replace natural micas. This was due, in part, to the inconsistency of the natural mica which could vary by season or geographical location of the mine. Synthetic versions of the mineral were developed as well as other substrates such as borosilicate (glass) or other metals such as aluminum, brass or iron.

Metallescent Pigments

Metallescent pigments are pigments that consist of a mica or other type of metal substrate coated with a metal oxide or a combination of titanium and other metal oxides such as iron or silica. These hybrid products combine the soft luster of the pearlescent pigments along with the higher reflectivity of the metallic pigments and are available in pearlescent shades of copper, gold, russet, bronze and silvers. Pigments that have an etalon structure of silica and titanium on their surface yield an unusual multicolor shifting product that changes on viewing angle. One pigment may yield as many as three or four different colors and are sometimes referred to as goniochromatic pigments.

Photoluminescent Colorants

Photoluminescent colorants emit light after absorbing energy (photons) from an external source. Photoluminescent colorants are in the form of both pigments and dyes and are broadly classified into fluorescent and luminescent categories.

Fluorescent Colorants

Fluorescent colorants are sometimes referred to as neon colors due to their emission of a brightly colored glowing hue similar to charged neon gases. Fluorescence only occurs in the presence of an appropriate stimulating light source. Biofluorescence is a naturally occurring phenomena that are due to proteins in living plants and animals that absorb photons and emits a light at a lower energy state. Fluorescent shades are generally in hues of red, green, yellow and blue violet. Optical brighteners (or bluing agents) fall into this category and are sometimes added to textiles, paints and plastics to offset the natural yellow shades that are inherent in the raw materials.

Luminescent Colorants

There are several different classifications of luminescence. Some examples of these types of luminescence are explained below. Chemiluminescence is created by a chemical reaction and can either biological in origin (biochemiluminescence) or electrochemical in origin (electrochemiluminescence). Electroluminescence is created when a current is passed through a material. LED lights are an example of electroluminescence. Crystalloluminescence is produced by crystallization. Fluorescent colors have a fast and immediate reaction to their exposure to energy whereas luminescent colors have a delayed and slowed reaction due to exposure to the energy source and the luminescent phenomena can appear to glow for several hours after initial exposure. This phenomenon is also known as persistent luminescence. Chemiluminescent materials are those that are driven by a chemical reaction rather than a UV or visible light reaction. An example would be a glow stick where a chemical reaction between two substances causes the electrons to enter an excited state and that energy stimulates a fluorophore that creates a sustained fluorescent glow.

Phosphorescent Colorants

Phosphorescent colorants are also stimulated by UV or visible light exposure rather than a chemical catalyst but the return to the normal state of the electron is a slower and longer-lasting glow rather than the immediate glow created by fluorescent colorants. Some examples phosphorescent-containing materials are the hands of watches and clocks, safety signs and other glow-in-the-dark objects.

Thermochromic and Photochromic Colorants

Thermochromic and photochromic colorants change color when exposed to heat (thermochromic) or a light source (photochromic). An example of an application for photochromic dyes is tinted lenses for eyeglasses that photodarken when exposed to sunlight and return to colorless when the user returns indoors. An example of an application for thermochromic dyes is flat thermometers and food packaging that indicate that a product is at the correct temperature for consumption.

Further Reading

https://en.wikipedia.org/wiki/Cornelis_Drebbel
http://www.scs.illinois.edu/~mainzv/HIST/bulletin_open_access/v33-2/v33-2%20p61-67.pdf Prussian blue
http://www.straw.com/sig/dyehist.html
http://colorusage.arc.nasa.gov/lum_and_chrom.php
http://www.naturalpigments.com/lac-dye.html
https://www.boundless.com/physics/textbooks/boundless-physics-textbook/introduction-to-quantum-physics-28/applications-of-quantum-mechanics-183/fluorescence-and-phosphorescence-676-4913/
https://en.wikipedia.org/wiki/Light-emitting_diode
https://en.wikipedia.org/wiki/Luminescence
https://en.wikipedia.org/wiki/Persistent_luminescence
https://en.wikipedia.org/wiki/Rubia
http://www.pbm.com/~lindahl/articles/food_coloring_agents.html
https://en.wikipedia.org/wiki/OLED#Polymer_light-emitting_diodes
http://timesofindia.indiatimes.com/city/ahmedabad/Chintz-dyeing-heritage-spins-comeback-yarn/articleshow/49973821.cms
http://www.chemeurope.com/en/encyclopedia/Kermes_%28dye%29.html
http://www.nndb.com/people/219/000099919/
http://www.historyofjeans.com/jeans-history/history-of-dungaree-fabric/

37 Color Measurement

Color Measurement is the quantitative expression of color and as indicated previously, there are a number of methods to quantify color. The best method of determining color is with the human eye. As mentioned previously, some humans see colors more distinctly and everyone sees and perceives color differently. However, most color scientists, engineers and manufacturers need to attach numbers to their color assessments so there are many different types of tools available to identify, quantify and differentiate colors. These analytical tools can help to provide direction in color matching, quality control in manufacturing and to develop new colors and color spaces. It also helps color professionals and designers talk to each other about color in a universal language. There are two basic types of technology for the measurement and classification of color: colorimetry and spectrophotometry.

Colorimetery

Colorimetry is the science of correlating color to human perception. Typically a tristimulus colorimeter is used to measure light and correlate it to the X (red), Y (Green/Yellow) and Z (Blue) components of CIE 1931 color space. Another approach to correlate color to human vision on the L*a*b* color space. The L* value relates to the lightness of a color using a numerical range of 0 – 100 where the darkest color (or blackness) is equal to zero and the lightest (or whitest) color is equal to 100. The a* values are represented as the red to green range where red colors are the positive values and green colors are the negative values. The b* values are represented as the blue to yellow color range where yellow colors are the positive values and blue colors are the negative values. These a* and b* values are plotted on an XY axis with the L* value piercing the center of the plot as a three-dimensional space. When a* and b* values are measured at zero, the color indicated is a true neutral falling right in the center of the red to green space or the blue to yellow space.

Color Trends and Selection for Product Design.
DOI: http://dx.doi.org/10.1016/B978-0-323-39395-9.00037-2

Spectrophotometry

A spectrophotomer is an analytical instrument that is used to identify and quantify colors. A typical spectrophotometer consists of two parts: a source (spectrometer) that produces light at specified wavelengths and a detector (photometer) that measures the intensity and amount of that color present at the specified wavelength. A traditional spectrophotomer is used to identify dyes that are dissolved into a solution and placed in rectangular cuvette inside of the instrument where the spectrometer transmits light of a specified wavelength that shines through the dye and then the photometer detects the signal and registers it as a voltage signal that is transmitted to a display device (galvanometer) and is printed as a spectral curve that is representative of the amount of energy created at different wavelengths.

For color identification and quantification, the spectrophotometer measures in the visible (400 to 700 nm) and sometimes the UV range (185–400 nm) where the human eye can detect color.

Spectrophotometers can also measure color from opaque surfaces such as plastics, paints, textiles and inks. In this case, the spectrophotometer measures how much energy is reflected off the surface of an object at specified wavelengths to identify and characterize colors.

Color Temperature Meter

The color temperature of white light is important for many industries including the film and television industries where cameras are calibrated for to achieve accurate skin tones. Initially, cameras are calibrated for white balance by focusing the camera on a white in ambient lighting and then adjusting the settings to achieve an accurate whiteness. This calibration will also provide the basis for all other colors in that light source. The basis of color temperature measurement started with William Kelvin in 1848 when he heated a block of carbon and noted the color changes as the temperatures increased. Initial colors detected were a slight glowing red that changed to yellow and finally reached blue shade white at its highest heat. This is the basis for the Kelvin color temperature scale. Correlated color temperature is the relative color of white light and is expressed in Kelvin temperature units. Warmer colors are indicated as yellows and golds whereas cooler colors are indicated as blues and whites. However, blues and whites are in the

7000K range whereas the reds and yellows are in the 2000–3000K range but human nature associates blues and whites as icy and cool and reds and yellows as warm and fiery.

Haze/Gloss Measurement

Haze and gloss meters do not measure RGB color but rather the surface of the colored part but rather measure such things as distinctness of image (DOI), haze, gloss and orange peel. Depending on the finish of a part, the appearance of a color can be dramatically altered. As an example, a glossy surface can yield a much more intense color where a matte or highly textured color can appear much more subdued. This is most useful for painted or surface finished parts.

Gloss is defined as the amount of light collected from an object at the specular angle. When a light source, the sun, an overhead light, a flashlight, contacts the surface of an object, one of four things can happen to that light energy. It can pass right through it (transmission), it can be completely absorbed by it and reflect back color, it can partially absorb into the surface and bounce back at many different angles (diffuse reflection) or it can reflect at the same but opposite angle it was illuminated (specular reflection). This specular reflection where the angle of incidence is equal to the angle of reflection is the true definition of gloss and at 100% reflection is equivalent to a mirror finish.

DOI is a measurement of how sharply an image appears in the reflection of a measured surface. If a highly polished surface is desired, a manufacturer will put a chrome finish on the inside of a plastic mold to achieve a mirror finish. In painted applications, a clear coat is often added to create a higher gloss surface finish. A mirror finish is defined as 100% reflection of the light that is subjected to a surface is collected or reflected at one angle of measurement. In other words, the angle of light incidence is equal to the angle of reflection.

Haze is defined as a milky surface of a gloss application and is frequently characterized as a halo effect and results from diffuse reflection of a light source. Luster or pearlescence is also a form of diffuse reflection where the light source transmits partially into the object and is reflected back the pearlescent pigments at different depths and at different angles from within the piece or the coating. Another definition of luster is the sum of light that is detected at the specular and diffuse angle combined. This is also known as contrast gloss. This combination of light reflection is the answer for why pearlescent pigments yield a three-dimensional effect in a two-dimensional piece or coating.

Orange peel is defined as a surface defect that can occur from poor weathering or poor application of the coating to the surface of a finished part. It is characterized by a subtle bumpy surface, similar to the surface of an orange. A surface that exhibits orange peel will decrease gloss and DOI readings.

Densitometery

A densitometer (spectroradiometer) measures how much light is transmitted through the surface of a product to quantify the spectral density of a colorant. This is particularly important for thin-walled plastics where a certain color might be achievable only if the part becomes more opaque.

Further Reading

http://www.aces.edu/dept/fisheries/education/pond_to_plate/documents/ExplanationoftheLABColorSpace.pdf
http://seesmartled.com/kb/choosing_color_temperature/
http://industrial.datacolor.com/wp-content/uploads/datacolorproductliterature/Instruments%20for%20Color%20Measurement_72.pdf
https://en.wikipedia.org/wiki/Xenon_arc_lamp
http://www.lrc.rpi.edu/education/learning/terminology/cct.asp
http://www.todayinsci.com/K/Kelvin_Lord/Kelvin_Lord.htm
https://en.wikipedia.org/wiki/Distinctness_of_image
https://en.wikipedia.org/wiki/Orange_peel_(effect)
https://en.wikipedia.org/wiki/Gloss_(optics)
http://www.minerals.net/resource/property/Luster.aspx
http://www.physicsclassroom.com/Class/refln/u13l1d.cfm

38 Color Matching

As mentioned previously, the most accurate color matching is obtained by the human eye. Unfortunately, the untrained human eye can tell us little more than two colors do not match. The trained eye of a skilled color matcher can generally detect differences in hue, chroma and lightness at different angles under different lighting conditions (metamerism). Most people see color differently and not everyone can detect subtle differences in colors. If two manufacturers are looking at two different parts in two different parts of the world, at different times of the day, the same color can appear dramatically different. Ideally, it is best to have the color manufacturer, the designer, the engineer and the manufacturer all in the same room, looking at the same part, in a light booth at the same time to agree on a color. This is difficult to arrange in fast-paced projects. As a result, analytical measurements from a colorimeter or a spectrophotometer are frequently relied upon to indicate color differences and establish a path for correction. These digital files can be transmitted to customers and suppliers in different locations so an agreement can be reached about the actual color target and differences.

The human eye can also be misleading. An example is when an automotive manufacturer called a supplier and said that the color they supplied to create an interior textured piece was off-color. When the supplier and the manufacturer looked at the piece in the automobile, it did appear to be off-color. However, when the supplier measured the piece with a hand-held spectrometer and compared it to the standard, the measurements were the same. After a bit of investigation, it was determined that the texture in the mold where the piece had been created had worn down and as the texture decreased, the appearance of the color shifted in the final part. When the mold had been refurbished, the desired color was restored.

Color Matching Versus Color Styling

Color matching is quite different from color styling. In most cases, color matching involves copying a color from a known standard or target.

Color Trends and Selection for Product Design.
DOI: http://dx.doi.org/10.1016/B978-0-323-39395-9.00038-4

Color styling involves the interpretation of a color into a specific medium. A similar comparison may be drawn to a photograph and a painting where a photograph provides an exact replication (in most cases) of a subject whereas a painting is an interpretation of a subject. A common mistake among designers is that once they choose a color for a project they ask the manufacturer for an exact match. Many times the manufacturer must match a painted target or textile and reinterpret this color into a molded plastic piece (or the other way around). If the designer will allow the manufacturer to interpret the color and provide a few different examples of color stylings then much time can be saved in going back and forth to achieve the perfect match which ultimately might be unattainable. Color styling also is frequently used to create and interpret color trend pallets. By allowing the color matcher the freedom of interpretation over an exact match, they are able to account for geographical differences in colors as well as the availability of raw materials for those regions. Of course, there is also a need for true color matches that must completely replicate the color of the desired target. In most cases, exact color matches are a drop-in for an existing product. Sometimes a customer would like to reduce the cost of their product by switching from a painted product to a molded-in-color product. These types of matches are difficult but not impossible to achieve in some cases but the disadvantage is that the color matcher may spend weeks or months to create this match that might not be achievable or the end product might be so costly that it saves the customer very little money to switch from one method of coloration to another. An easier color match to achieve is when a customer asks for a plastic to plastic match or a paint to paint match. Since the color matcher is working in the same medium, it is easier to achieve the match. However, this is usually a cost-savings initiative and the customer is shopping around for a cheaper price to color their product. There is no guarantee that the color matcher will win the business even if the color match is achieved since it might come down to the price of their product versus the incumbent. However, if an exact color match is desired, there are a number of questions that the color matcher must ask of the designer or engineer. There is a communication gap that exists between many technical and creative team members. It is important to bridge that gap whenever possible.

Further Reading

http://www.colorbasics.com/Metamerism/
http://hyperphysics.phy-astr.gsu.edu/hbase/vision/cieprim.html

39 Spanning the Communication Gap Between Technical/Commercial Suppliers and Creative Clients

As a designer, it is important to be as specific as possible about the color that is desired. One of the best methods to ensure success is to provide a physical target for the match that the supplier can keep or borrow. If possible, it is best to provide the color target as close to the final application. As an example, provide a painted panel if the final application is for an automotive paint application. If the final product is a cap for a detergent bottle, it would be best to provide a target that is injection molded with a similar shape and thickness, in the same polymer if possible. This usually does not happen. Frequently, the client will provide a Pantone number and have vague information about the polymer or vehicle. Obviously, this will make providing the correct match much more difficult. Sometimes designers will go into great detail about what they do not want rather than what they do want. Some of this helpful but too much will create confusion. Designers also tend to discuss colors in more artistic or cultural terms (design speak). Be sure to ask for clarification if the discussion becomes too technical.

As a technical or commercial person, be sure to limit your use of acronyms that designers might not understand (tech speak). If you need to discuss plasticating methods, paint vehicles, ink curing methods or color measurement, check with your client to make sure they are following you. Most designers are eager to learn about how their products will be developed but can quickly shut down if a canned technical or commercial spiel does not specifically relate to their project. Here are a number of questions that should be asked and answered in a meeting for a new color project:

What Is the Purpose of the Match?

This first question can tell the color matcher a lot about how much time they would like to dedicate to the project or if they want to take

Color Trends and Selection for Product Design.
DOI: http://dx.doi.org/10.1016/B978-0-323-39395-9.00039-6

on the project at all. Frequently, customers are quite vague about this question and it pays to get a firm answer. If they are just shopping for a better price or are unhappy with their current supplier then it pays to probe about the details of the impaired relationship. It is possible that the new supplier will not be able to meet the expectations of the customer if the customer's demands are unrealistic. At this point, it is a good idea for the color matcher to vet the customer to determine if it is worth their while to pursue the match. It is also not unreasonable for the color matcher to charge for their services but this is not a common practice.

Will You Specify Our Product if the Match Is Acceptable?

Some brand-owners will "allow" suppliers to color match an existing product or a new project but are unwilling to specify their color products to the molder or convertor. Their rationale is that at least they are giving the convertor a chance to look at their product and consider buying them if the convertor's current supplier cannot match the color. This does happen occasionally where the convertor will contact the color matcher to purchase the material but a more likely scenario is that the convertor will send over the color match to their current supplier and have them match it. It is best to charge a fee for the initial color development work in this case if possible to cover the development colors of the match. It also happens that a color matcher might create a new color that the brand-owner or specifier might like better than the color provided by the incumbent and then force the convertor to use the new color match. This is a rare occurrence. This could be a good opportunity for the color matcher to ask for payment of services. Generally speaking, color match services are performed free of charge for good customers or new customers that are willing to commit to specifying the color from that match if the color is deemed as acceptable.

Is This a New Project or a Refresh?

The value in asking this question at the initial design consultation is that it will give the supplier some key information about the scope of the project. If it is a new project, than the development times can be quite long (2–3 years) until production. If it is a product refresh of an existing line of products or a part then turnaround can be quite rapid. Also for a new project, the color approval process is longer and might not be decided by the design team but rather by the marketing team or the product team that is sponsoring the program. Usually for a refreshed color project, the cycle is much faster and decisions can be made directly by the designer or engineer.

Are There Other Products or Colors in the Project?

Sometimes a client is only looking for one new color or a refreshed color that is part of a series or that must harmonize with other parts. As an example, if the client is asking for an off-white color for a polypropylene cap, then it is important to know the color of the container for which the cap will be used. If the container is a blue-shade white, than the color of the cap should also be a blue-shade white rather than a yellow shade white that will clash with the container. Also if one part of the series contains effect pigments and the designer is requesting a color that contains a metallic effect, it is important to understand what type of metallic effect is used in the other products. A white pearl effect will appear radically different than an aluminum flake effect even if the base color is the same.

What Is the Final Product?

It is important to understand the application for the color. Will the product be exposed to the outdoors? Will it come in contact with food? Is it for children's toys? Where will the product be sold? Is it for a pharma or drug application? Anytime the customer answers yes to any of these questions, more questions should arise from the color matcher. As an example, if it is a color that is to be used for outdoor furniture, a light-stable colorant or a UV-stabilizer might be needed. However, it is best left to the customer to decide if they want a color match that contains a UV-stabilizer. Even though medical devices are manufactured in clean-rooms, it might not be necessary to use pharma grade raw materials depending on the customer's needs. The same is true for food packaging. Generally, if you ask a customer if they want you to use FDA-approved materials or lightfast colorants they will almost always say yes even though they might not need it for their application so it is better for the color matcher to ask about the application first and then follow up with regulatory requirements since these colorants and additives can dramatically add to the cost.

Is the Match Cost-Sensitive?

Perhaps a better question is how Cost-Sensitive is the Match since almost all color work has some cost-sensitivity associated with it. Color matchers have a wide array of colorants available to use for a match. Many color manufacturers offer similar colors to their competitors and prices vary widely. If the project is extremely cost-sensitive, a color matcher might opt to use a less-expensive product that lacks some of the chemical or regulatory characteristics of a higher-priced product that yields an identical color in a match. If the color matcher can look at the target and see that high-price colorant will be needed to achieve the match and there is no less-expensive colorant available, they may decide not to match the target.

What Is the Substrate?

It is important to know the composition of the material that the colorants will be applied onto or into. This will dramatically affect how the colorants will interact with that product and if the color will change when added to the finished product. When preparing a match for a plastic product, it is important to know how the product will be manufactured. Will it be a blow molded bottle? An injection molded toothbrush? An extruded film that will be thermoformed or inset molded into another piece? If it is a blow molded bottle, is it a mono-layered or a multilayered bottle? Will it be molded by extrusion blow molder or stretch blow molded? How many layers would the multilayer bottle have? Would the color go in the outer layer (most likely) or would there be a clear layer on top of that? What polymers will be used in the layers? Will there be fillers or regrind in the inner layers? Will the fillers or regrind add to the color of the bottle? What will be the thickness of the entire bottle? What will be the thickness or ratio of each layer (typically it is an 80/10/10 or a 90/5/5). Not only is it important to determine the type of polymer used in a product, but it is best to determine the brand and grade of the polymer. If possible, the match should be created in that exact polymer since some polymers are tinted or have other additives that can interact with the colorants.

What Are the Dimensions of the Finished Product?

Ideally a color match should be presented to the customer in a similar shape and size as the finished product. Most automotive customers are happy to look at a match in an injection molded plaque or a chip for plastics applications but the best suppliers offer these chips or plaques that are textured with the same set of textures that are common for that customer. Similarly, for packaging customers, they might be satisfied with an injection molded chip or an extruded piece of film but the best suppliers can create color matches in mono or multilayered bottles or can injection mold a cap that can give the customer a much better idea of how the color will appear in the finished product. Having sample types close to a customer's final product frequently separates serious color suppliers from the rest of their competitors.

What Is the Surface Finish?

Knowing whether the surface of the product will be textured, smooth or polished is very important in providing a correct color match since colors will shift depending on the finish. Also if the client is trying to achieve a soft-lustrous effect or a color-shifting effect then these types of effects work best with a very smooth surface finish. If the surface is a matte finish or even highly textured then the effect pigments must be added at a higher concentration in order for the special effect to visible. In some instances, effect pigments cannot be seen if the surface finish is highly disrupted or textured. As a general standard, surface finish ranges from rough and highly textured to matte to polished to chromed. These terms are quite generic and relate to the surface finish of the mold that is used to create a molded plastic piece. A highly textured mold might create plastic pieces that have the appearance of a rough, raised or highly etched pattern. A matte finish on the mold might yield a slightly buffed or scratched appearance. A polished finish would have the appearance of being quite smooth with little to no surface defects.

A chromed mold usually is a highly polished mold that is subsequently electroplated with a chrome surface that can yield an almost mirrored, high gloss finish on the surface of the plastic piece. Generally speaking, the more smooth the surface on the mold, the more expensive it is to manufacture the mold. Conversely, a mold that is created with a specific grain or pattern can also be quite expensive.

Coating or Molded-In Color?

This question is usually answered by looking at the target and seeing if the final product is paint, plastic, ink or textile. Although plastic matches are mostly molded-in-color, there are occasions when plastic may be coated with a paint or ink. Molded-in-color means that color is added to the entire piece or in the case of a multilayered bottle or multilayered film, the color might just be in one or two layers. The same is true for textiles where dyes might be added directly to fibers before spinning or they might just be applied to the surface as a decorative coating.

How Will the Product be Packaged?

Even if the supplier is providing color for a packaging application, the package might be encased in a shrink wrap or even in an outside box. The color and composition of this outer packaging can influence not only the appearance of the finished product but can also affect the stability of the finished product. The specifier might only think about light stability on the shelf but not take into consideration what could happen to the product when it is contained within a secondary package.

What Is the Light Source?

It is important to understand what type of lighting will be used to evaluate the proposed color sample as well as where the product will be sold or reside. Many color submissions are evaluated immediately in a conference room or in an office that is illuminated with fluorescent lighting that tends to cast a blue hue on the submission. Incandescent light tends to make a color more yellow. However, the finished product may end upon a retail shelf that may be in direct sunlight for part of the day or under LED lights (that tend to be more white) for another part of the day. Hue can shift depending on the light source or even the angle of the light. It is best to qualify colors in a calibrated light booth that contains a variety of light sources that can be viewed at different angles.

Some of the more common light sources include Daylight (D50, D65 or D70), Incandescent A (150 W), Horizon (2300K), fluorescent (cool white, U30, U35, TL83, TL84) and ultraviolet. Frequently, the color will shift when the light source or the texture has changed. This is called metameric failure. The simplest definition of metamerism is when two different color samples appear (by eye) to match under one light source but then appear to not match under a different light source.

Incandescent light is created when an electric current is passed through a thin metal filament (usually tungsten) and the resulting glow usually emits a cool yellowish cast. General service bulbs are either frosted or clear and emit light equally in all directions. In contrast, reflector bulbs have a mirror coating on the inside of the bulb and focus all the emitted light in one direction. This generally doubles the amount of light that is focused at one angle. Tungsten-Halogen bulbs are also incandescent bulbs but yield a much bluer cast than a single tungsten source.

Light-emitting diodes (LEDs) create visible light when a current is passed through an inorganic semiconductor. Emitted colors are yellow, blue, red and green. White light is created when all of these colors are combined and exposed to excited phosphorous which reduces the intensity of the colors. This is not considered to be a pure white light and can vary in shade depending on the ratios of the individual components. Other types of LEDs are organic light-emitting diodes (OLEDs) and light-emitting polymers (LEPs). OLEDs contain an organic semiconductor that acts as the emissive illuminant. These are typically used as display screens in smart phones and computer screens and are capable of producing very deep colors and rich shades of black with high-contrast ratios. LEPs are polymer light-emitting diodes that consist of an electroluminescent conductive polymer that emits light when exposed to an electrical current. Different conductive polymers emit different colors of light. Although these types of lighting are generally not used to view color submissions, they will influence other light sources when they are present. OLEDs and LEPs can be present in clothing or textiles, television screens and architectural glass panels.

High Intensity Discharge Lamps produce light when an electrical current creates an arc between to cathodes that are deposited with metallic salts that vaporize upon ignition. Low pressure sodium vapor lights are a type of metal halide lamp that emits an orange or reddish light. High pressure sodium vapor lights have a closer to white emission but still cast a pinkish orange glow. Ceramic halide lights create a broader spectrum of light that are more neutral white light. These lamps are frequently used for indoor greenhouses and aquariums.

Xenon lamps create light by sending an electrical charge through an ionized Xenon gas. The resulting emission is a bright white light that is a multispectrum similar to sunlight. They are frequently used in projectors, search lights and weather-testing equipment.

How Close Does the Match Need to be?

To some specifiers, white is white and subtle variations in shade that shift from yellow to blue are inconsequential to their project. Other specifiers are looking for a very specific hue or might even request a certain level of chromaticity or brightness. Frequently, a specifier or a designer will give a color matcher a target and want it to match exactly. The color matcher will need to decide how close the match needs to be based on the amount of time allotted for the match, the types of raw materials available, the cost of the raw materials and the final application for a match. If the color matcher is given a painted or coated color target and is asked to create the same color and/or effect in an injection molded cap or a blow molded bottle, the color match will be an approximation at best. Ideally, a color matcher would like to have a target that is in the same medium (paint to paint or plastic to plastic), an infinite amount of time to create the match, and an unlimited supply and budget for raw materials. This almost never happens. Also what appears to be a match for one set of eyes can be radically different for another set of eyes. Using a spectrophotometer to measure color differences can help level that playing field but still might not meet the wishes of the specifier and it is best when the color matcher and the specifier can view the matches together at the same time under the same lighting conditions and they can discuss the differences or similarities that they see in the color. This also rarely occurs.

How Soon Is a Match Needed?

The answer to this question is almost always the same: as soon as possible. But a good color matcher usually has a stack of projects to complete in the lab so then the problem becomes what to complete first and for whom. A better question would be to determine the scope of a project: when will the colors be reviewed and selected, when will the production trials start and when will the color go into production. Also it is important to vet the opportunity with questions like: where the product be produced and what will be the proposed quantities. If the product is being manufactured in a different country, it will be important to know what raw materials can be used in that country and if they are available to manufacture in that country. If the volumes of the product are too small or the color matcher does not have manufacturing capabilities in that region, it might be best to leave the project for someone else. It is difficult to arrive at the color match on the first submission. There can be several submissions before the correct color is achieved. It is also important to ask if they will need a small amount of material that they can test in their manufacturing with the match. Some clients require testing material with the color submission but it is better to wait until the color is approved before providing a sample in the event that the color needs some correction prior to trial. Also, some smaller manufacturers will not need a large amount of color to complete their production run so if the supplier provides a larger free sample, they might miss out on a sale.

Further Reading

http://www.energystar.gov/products/lighting_fans/light_bulbs/learn_about_led_bulbs
https://en.wikipedia.org/wiki/Metal-halide_lamp
http://www.americanlightingassoc.com/Lighting-Fundamentals/Light-Sources-Light-Bulbs.aspx
https://en.wikipedia.org/wiki/Xenon_arc_lamp
http://www.chemeurope.com/en/encyclopedia/Pigment.html
http://www.appropedia.org/Blown_film_extrusion
http://www.bpf.co.uk/Plastipedia/Processes/Blown_Film.aspx
http://www.macroeng.com/fundamentals-of-cast-film-extrusion-technology.php

40 Coloring of Plastics

Colorants, Pigments and Dyes

Any additive that contributes color to a material is considered to be a colorant. A pigment is an insoluble additive that changes the reflected or transmitted light through selective absorbance of a material. A dye is also a colorant that is soluble additive that changes the color of a material. There are many books that have been written on the coloring of plastics so this section will provide an overview of color processes and applications.

Masterbatch Versus Precolor Versus Concentrate Versus Dry and Liquid Colors

A masterbatch is a plastics formulation that contains additives that are essential for the appearance and performance of the final product. A masterbatch can also include colorants. Generally speaking, a masterbatch is comprised of plastic pellets that contain concentrated amounts of additives such as processing aids, colorants, antistats, light stabilizers, fillers, etc. There are many different types of masterbatch vehicles including plastic pellets that are customized to be compatible with the final polymer used in the product. The vehicle can also take the form of powders or even liquids that are compatible with the end use polymers.

A color concentrate is usually defined similarly to a masterbatch except it consists only of colorants rather than additional additives that would be found in a masterbatch. Both color concentrates are added or "let down" into the final polymer prior to molding or extrusion into the final product. This blending process is usually a low shear process and can occur in a bucket or a tumbler at the molding machine. The amount of functional ingredients and colorants varies widely but are usually in the range of 0.5% to 5.0%. This is sometimes indicated as a let-down ratio which is a ratio of natural material to the concentrate product. As a rule of thumb, the higher the concentration, the more expensive the product.

Color Trends and Selection for Product Design.
DOI: http://dx.doi.org/10.1016/B978-0-323-39395-9.00040-2

A precolor or compounded color is a product that is not a concentrate but contains the total amount of colorants and other additives that will be used in the final product. Some molders prefer compounded materials because they do not need to mix the materials prior to processing or because the dispersion of additives and colorants can be more uniform. This is particularly true with dyes when they are formulated in very light colors and low concentrations. If these dyes are not properly dispersed, it may cause color streaks or agglomerates (clumps) in the final product.

Liquid colors are usually colorants dispersed in oils that are compatible with the final media. These liquid colors are usually quite viscous and require specialized delivery systems or pumps to add the colors directly into the molding or extrusion equipment. Liquid colors can be extremely high concentrates of colorants and are generally seen as very economical after the initial purpose of the additional delivery systems required. This type of color is best for high-volume application for manufacturers who continuously create the same part in the same color. However, if the customer makes frequent color or design changes, there can be a great deal of waste when making these change outs. Since the colors are an oily liquid, they can cause a great deal of housekeeping and maintenance when compared to masterbatch or other types of pelletized colors.

In contrast, dry colors are best for small runs and quick change overs because they are simply blended with the vehicle prior to processing. Dry color also yields consistent coloration as well as good dispersion.

Coloring Films

Films are created by different methods of extrusion. Blown films and cast and extruded films are the three most common methods of creating films.

Blown films, also commonly known as tubular films, are created by extruding plastic through a tubular-shaped die and then stretched and blown with air to create a very thin-walled film. Polyethylenes such as LDPE, LLDPE and HDPE polymers are common resins that are used to create blown films. The most common product created by blown films is disposable grocery bags and other types of food and consumer packaging. Other types of film include greenhouse films and shrink wrap films. Blown film machines are typically very tall vertical towers that stretch and blow the films into their final form. After the air exits and the film cools, it is rolled through a series of rollers that flatten it into layers. Blown films have very good mechanical properties because it is stretched and blown both horizontally and vertically during extrusion. However, film thickness is much more difficult to control when compared to extruded or cast films.

Cast films are created by a similar extrusion method where the polymer pellets or powder is mixed and melted as it travels along the flights of an externally heated screw. When the molten polymer reaches the end of the screw, it is forced through a flattened die. This die can split the polymer or polymers into many different layers. If different types of polymers or polymers with different colorants and additives are extruded into one film, then multiple extruders can be fed into one die.

In recent years, there has been an emphasis on light-weighting cars, to increase fuel efficiency. An obvious strategy is to replace metal components with plastic equivalents that are considerably lighter and easier to modify the design. However, coloring these pieces is somewhat of a challenge. Automotive manufacturers favor traditional painting methods to color steel and have tried to implement that process onto plastic components. Painting plastics represents a unique set of challenges and the traditional types of paints that are used on steel are not appropriate for all types of plastics. Also, if the painting process is not perfect, the paint is not as easily removed from a plastic piece as it is from a metal piece and the scrape rate is too high to be cost-effective. Another strategy was to try to match paint finishes with molded-in-color, injection molded pieces but were not able to match the finish or dimension of paints. Extruded, cast films were developed to create the same colors and effects of automotive paints that can be laminated onto the surface of plastic components. Automotive films that require very high quality class A surface finishes, use this multilayer extrusion process where the top layer is a high-gloss clear layer (similar to a clear coat in automotive exterior paints), the next layer might consist of semitransparent effect pigments. A mid layer may contain absorption colorants that give the film its overall color. There can be several layers underneath these color and effect layers that contain other additives such as tie layers (to join the multiple layers together), adhesives, antistatics, UV stabilizers, etc. There are also other types of multilayer films created by this method that yield an iridescent or rainbow effect. This effect is achieved by extruding several different polymers of different indices of refraction and combining them into one film. A current trend in automotive colors and finishes is moving away from these high-gloss finishes and a more matte effect is becoming more popular for automotive surfaces. This will increase opportunities for more molded-in-colors and more economical methods of plastics coloration.

Flexible and Rigid Packaging and Containers

As plastics technologies and processing improves, the line between flexible and rigid packaging becomes less defined. Generally, rigid packaging is defined as a container whose shape is not unduly influenced by its contents. Rigid packaging is frequently used to protect its contents from stress and breakage and is easier to stack and store. Blow molded shampoo bottles and milk jugs are common examples of rigid packaging. Injection molded storage containers and rotational molded fuel tanks are also considered to be rigid containers. Flexible packaging and containers are broadly defined as those components that take the shape of their components or surroundings. Pouches and pinch tubes are frequently used in cosmetic and food packaging and can also consist of many layers containing polymers and additives that possess barrier, and photostability properties as well as increase shelf-life. Flexible packaging is sometimes considered, "greener" packaging because it is a more efficient use of space and can reduce the overall carbon footprint of a brand or product.

Further Reading

http://www.energystar.gov/products/lighting_fans/light_bulbs/learn_about_led_bulbs
https://en.wikipedia.org/wiki/Metal-halide_lamp
http://www.americanlightingassoc.com/Lighting-Fundamentals/Light-Sources-Light-Bulbs.aspx
https://en.wikipedia.org/wiki/Xenon_arc_lamp
http://www.chemeurope.com/en/encyclopedia/Pigment.html
http://www.appropedia.org/Blown_film_extrusion
http://www.bpf.co.uk/Plastipedia/Processes/Blown_Film.aspx
http://www.macroeng.com/fundamentals-of-cast-film-extrusion-technology.php
Charvat, R., 2003. The Coloring of Plastics, second ed., Wiley-Interscience, Hoboken, NJ.

41 Coloring Labels and Color Printing

A recent trend in packaging has been to increase label size to cover most of the package and to color only the label and not the package itself. This is an extremely cost-efficient method to decorate a package or container and also allows for more flexibility for brands with multiple products to use the same container and covering it with a different label.

There are hundreds of different types of printing that have been developed since the beginning of civilization and vary widely in quality and speed of application. The first recorded prints were wood-block prints that were discovered as early as AD 200 in China. In the 5th century, the Chinese used a method of wood-block printing called, *Jia Xie* to create colored patterns and images on silks. This method used two carved wooden blocks that had plugged holes on the outer surface to pour in liquid colors or inks that would be pressed into the silk that was tightly folded between the two blocks. Currently, the most common types of printing are gravure, offset digital and 3D printing.

Gravure Printing

Gravure printing is a very high quality method of ink printing that requires the use of an engraved plate or roller. The multicolor gravure print was developed early in the 20th century. The multicolor process is created by printing the same image, multiple times using CMYK color process model. This process has an excellent contrast ratio of light to dark. It also a costly and time-consuming process to set up and is only feasible for high-volume prints such as high-end books, prints or fine art prints.

Offset Printing

Offset printing is a version of gravure printing where the image is transferred from an engraved plate and transferred (offset) to a flexible

Color Trends and Selection for Product Design.
DOI: http://dx.doi.org/10.1016/B978-0-323-39395-9.00041-4

web or sheet that has hydrophobic and hydrophilic sections that draw the ink to the image section and repel it from the background sections. This is a high-volume process where reels of paper are rolled across this web and continuously inked and printed. It is the highest quality printing method that is used for cost-sensitive applications. The quality is not as high as gravure printing which shows higher contrast ratios and sharper resolution when compared to offset printing.

Pad printing is a flexible version of offset printing where an etched plate, is inked (flooding), scrubbed and dried to remove residual ink and then the inked image is transferred (pick up) onto a flexible silicone pad that is subsequently pressed onto the label (ink deposit). The advantage of this method is that the printing can occur on a contoured or a curved surface.

Digital Printing

Digital printing is simply transferring a digital file or image file such as a jpeg, pdf, word, excel, photoshop, etc. that was created on a computer and then sent to a printer to create a physical print using CMYK ink. This eliminates the need for a printing plate and other secondary printing applications to create a printed article. The CMYK process model uses three primary colors of ink: cyan (blue), magenta (purple/red), yellow and key (black) to print colors. When printing these as solid colors, seven different colors are possible. Halftoning is a process where the ink is applied as much smaller dots on the surface, rather than a solid application, so that a more whitened color appears visible to the human eye and an infinite number of colors are possible in the printing process. As an example, an application of magenta can shift from deep purple to a light pink just by varying the amount of white space between the magenta dots. The smaller the dots, the more white space is visible and the result is a lighter color. The ink can be deposited onto a variety of substrates including paper, photographic paper, stone, textiles, glass, plastic, metal and other substances. Three common types of digital imaging are inkjet, dye sublimation and laser printers. These printing methods are usually more expensive (due to the ink and paper cost) than traditional offset methods, they also lack the resolution of gravure printing but are rapidly prepared and printed when compared to other methods.

The inkjet method uses heat to vaporize and disperse tiny droplets of ink onto the paper surface where it forms a thin film consisting of 0–8 droplets of the CMYK per dot, depending upon the required shade as indicated by the rendering software.

Dye sublimation printers are a heat transfer printing process used mostly for textiles. These printers use only sublimated dyes. Digital images are printed onto heat sensitive papers that are subsequently pressed and heated to transfer the image onto the textile. This yields a high resolution and permanent image on the surface of the textile that yields true and consistent color tones that rival a chemical photographic quality.

Most inks are made from dyes using water as a vehicle and these types of inks are quite vivid in color and suitable for short-term interior applications. However, they are also water soluble and not light fast so if the user needs archival or prints for exterior applications, solvent-borne inks that use pigments rather than dyes are a better choice. UV curable inks are another type of ink that uses UV light to rapidly cure or dry the ink. This is an interesting process because the rapid cross-linking process can be used to create disruption patterns with effect pigments on the printed surface. This is a much more expensive process than other drying methods but it is quite fast and can print on many different surfaces rather than costly specialty papers. It also yields high-intensity prints because it uses all of the colorant dispersed on the surface as well as being a much higher quality print. Dye sublimation inks consist of special dyes that can sublime (change from a solid phase to a gaseous phase back to a solid phase) without having to use a liquid carrier. This phase change is catalyzed by high heat and pressure and creates images that are quite vivid and stable.

Laser printers do not use ink but rather a charged toner that is electrostatically glued to a pattern that is temporarily lased onto the surface of the drum. The charged colorants (toner) stick to the lased pattern on the drum that is subsequently transferred onto the paper as a print. The paper is then run through heated rollers to melt the toner onto the paper when it becomes permanently affixed to the paper. Laser printing is much faster and less expensive than inkjet printing but color quality and resolution are quite poor in comparison. However, it is comparable in cost to offset printing but considerably faster for production work.

Three-Dimensional Printing

Three-dimensional printing (3D printing) started in the early 1980s in Japan. It is not a traditional ink printing process where a liquid ink is applied to an engraved surface and then pressed to a sheet or paper. There are many different types of 3D printing but the most common class of printers uses the additive layer manufacturing process. Typically, 3D printers are small desktop units that consist of a small heated extruder head that melts a thin stream of colored plastic that moves in an X/Y/Z motion directed by a digital .stl file from an attached computer. These thin layers of extruded plastic are temporarily "glued" onto the printer platform. The layers are then built on top of the previous layer until a three-dimensional shape is formed. The colored plastic is usually wound on a spool that is attached to the back or side of the printer and is fed into the extruder through a plastic guide tube. Although there are many types of thermoplastic resins used in 3D printing, two of the most common polymers used are PLA and ABS polymers. ABS is the more durable of the two polymers but has an unpleasant odor while printing. PLA is a green polymer and is much more transparent so it can be brightly colored. It is also more brittle than ABS and is not recommended for functional pieces that will endure stress or repeated impact. 3D printing is not limited to plastics. Recently, 3D printing has been used to create products out of chocolate, metallized plastics and wax. Printed wax figures can be used to create metal pieces using a lost wax process where the initial design is created as a CAD file that is printed in wax. This wax figure is then coated in ceramic and fired into a hard shell that is subsequently used to cast objects into metals. There are also 3D printers that do not extrude wax or plastic. Some of these printers extrude a liquid mixture of minerals such as gypsum that is suspended in a liquid adhesive, such as an acrylamide gel that hardens immediately upon deposition to the print bed. Other printers use liquid photopolymers that cure upon exposure to UV light. The great advantage to this method is the ability to mix and combine colors and patterns for each printed piece. The colors are added as liquid suspensions from cartridges inside the printer and function more like liquid inks in a digital printer. 3D printers are a great tool for designers and engineers to create unique three-dimensional examples of new designs and are used mostly to make prototypes and models rather than functional pieces because of their lack of dimensional stability when compared to other manufacturing methods. This is also a very slow process where a small plate or object (such as cellphone case) can take an hour or more to print; however, as this industry continues to evolve, the prints are getting larger, faster and more dimensionally stable.

Aquagraphics Printing

Water transfer printing or aquagraphics is a decorative where a pattern or a colored image is printed on a solvent-soluble web. This web is floated on the surface of a water bath and the three-dimensional object like a fender or an instrument panel is submerged into the bath and the graphic sticks to the object. The object is then dried and sometimes cured and posttreated with a protective coating and used in applications such as a carbon-fiber patterned gun stock or python-patterned motorcycle helmet.

Vehicle Wraps

Vehicle wraps are used to cover trucks, buses, cars and trains with decorative or informative printed media. These wraps are highly detailed patterns, colors and logos that are printed on large vinyl sheets that are adhered to the surfaces of the vehicles. The vinyl is backed with a unique adhesive that is comprised of glass beads that allows the print to be repositioned numerous times for exact placement until final heat and pressure are applied. In most cases, these wraps are applied to commercial vehicles for advertising and brand promotion; however, there has been a trend to use these wraps as a replacement for auto motive paints, especially in the case of municipal vehicles and livery cabs that will have resale value beyond their initial purpose. As an example, police cars are frequently painted with bright colors and graphics for visibility but when they are re-sold, they are repainted to remove their official identification. When a vehicle wrap is applied, the wrap is simply removed for the new owner revealing the original paint or primer color under the wrap. This is a much less expensive method of covering and decorating a vehicle when compared to custom automotive paints. However, these wraps are less durable and damage much easier than traditional automotive paints but these wraps are improving in durability as polymers and processes for application are evolving.

Further Reading

https://en.wikipedia.org/wiki/Resist_dyeing
http://www.3ders.org/3d-printing-materials.html
http://www.aquadipgraphics.com/

https://en.wikipedia.org/wiki/Offset_printing
http://www.padprinting.net/proc_about.html
https://en.wikipedia.org/wiki/Inkjet_printing
http://www.cantorfoundation.org/resources/the-lost-wax-casting-process/
http://www.photopolymer.com/
http://www.livescience.com/34551-3d-printing.html
https://en.wikipedia.org/wiki/Wrap_advertising

42 Green Colors and Plastics

When discussing green colors and products, the discussion is generally limited to those products that offer a smaller carbon footprint than traditional products. Green products might be further defined as plastics that are derived from a nonfossil fuel source such as PLA (polylactic acid) which is derived from corn. Most natural products derived from plants and animals are also considered green or earth-friendly although sometimes the processing of these natural products require copious amounts of water or energy and can release toxic fumes or vapors during production or in the waste water.

Bioplastics

PLA is known as a compostable plastic since it breaks down in a landfill in 40 days under normal heat and pressure. However, there is still a cost to society and to the environment in the production, use and decomposition of this product. Corn has long been regarded as a food source for animals and humans so growing corn that is used for a bio-polymer reduces the amount of corn or other plants that could be grown for animal or human consumption. Also high heat and pressure are required to process corn to convert it into plastic. This generally requires the use of fossil fuels in an industrial setting. When the PLA item is discarded and ends it life in a landfill, it will release greenhouse gases such as methane and CO_2. All green products come at a cost. Most designers and project engineers are looking for the easiest green products. In many cases, they are looking to replace an existing product with a greener product but need to retain the same physical properties, costs and similar processes to incorporate.

In terms of plastics, it commonly held that the development of all basic polymers started out as green polymers such as melamine and nitrocellulose but these were replaced by petroleum products that were more stable, durable and easier to process. As an example, nitro-cellulose (which is still used today in nail enamels) was first used to make billiard balls. Nitrocellulose is created from raw cellulose which

Color Trends and Selection for Product Design.
DOI: http://dx.doi.org/10.1016/B978-0-323-39395-9.00042-6

is a starting product for many cellulose-type of polymers. As nitrocellulose ages and dries, the polymer becomes explosive and can decompose on slight heating or impact. Most billiard balls today are made from cast polymers such as polyesters or acrylics. Another cellulose-derived product is cellulose nitrate that is quite similar to the polymer used in nail enamels but the other application is smokeless gunpowder. Cellulose Acetate Butyrate is another polymer derived from natural cellulose. This polymer is used as window caulking as well as the gum in most chewing gums confections.

Another natural "green" product that was used in the 20th century was asbestos. Asbestos is a mineral that was frequently used in floor tiles, wallpaper, adhesives, house insulation and in automotive applications. It was also combined with vinyl that acted as a binder for these applications. Asbestos is a natural material that is easily mined from the earth and processed. It has excellent functional properties such as heat resistance and wear; sound dampening and electrical resistance and was certainly considered to be an easy green material in its day. However, today the material is banned and is regarded as quite toxic because it can cause lung scarring and cancer in its airborne state. Asbestos floor tiles were replaced with linoleum and vinyl flooring.

Bakelite (phenolic formaldehyde) was a newer synthetically derived polymer that was created by combining phenol and formaldehyde under high heat and pressure. Although this thermosetting process made incredibly durable and stable molded pieces, they are not recyclable due to their inability to break the polymer chains when reheated. Many of the markets that Bakelite once dominated are now using ABS polymers as a replacement since it is a thermoplastic material that is more flexible and recyclable. There is a renewed interest in Bakelite since it has a coldness and a heft that is difficult to replicate in thermoplastic materials that tend to be lighter and warmer.

Melamine resin is another ceramic type of thermoset polymer that is commonly used in dishes, eating surfaces and whiteboards. It is a derivative of urea. Urea can be naturally derived or synthetically created and may be characterized as a green product in some industries. It is also a component of Pigment Yellow 150 used to color plastics. However, in 2007, melamine was added to infant formula and other food products for humans and animals in China to boost the detectable level of protein in the products. This lead to the deaths and illnesses of many babies and animals globally.

Linoleum is also a green product, made from linseed oil which is processed with cork and other fillers such as sawdust to create soft and sanitary natural products. A heavy gage linoleum was known as battleship linoleum was used to replace wooden decks on the flooring surfaces of large ships and destroyers but was discontinued after the Pearl Harbor attack when its high flammability characteristics were discovered. Most linoleum flooring was eventually replaced by the nongreen vinyl flooring which had its own liabilities including toxic plasticizers and its contribution to acid rain during processing.

PHA and PHB (polyhydroxyalkanoates and polyhydroxybutyrates) are biopolymers that are created from bacterial fermentation of fats or sugars. These linear polyesters are somewhat elastic and also have greater UV stability and are less hygroscopic that PLA materials. PHB is similar to industrially synthesized polypropylene with good moisture and barrier resistance for food packaging. These products have a generated a great deal of interest by the medical device and surgical manufacturing industries since they can be used for surgical sutures, stents and filling agents.

Recycling

Waste materials that end up in landfills and have no hope of reuse or recycling is a flaw of the initial product design process. As designers and engineers select materials for their products, there should be an emphasis on what will happen to that product when it is no longer useful for its original purpose. It is important to understand that recycled materials might have compromised physical qualities when compared to new or virgin materials so care must be exercised to understand that these recycled materials might not meet the same rigorous physical testing or regulatory requirements than their virgin counterparts. From a design perspective, another alternative might be to make a product that can be reused or repurposed as another product.

Rainbow recycling is a grassroots organization that is trying to create a consistent color scheme for recycling bins and labeling. Red is for metals since metals need to be heated red hot to melt. Blue is for plastics and their association with water bottles. Green is for paper since it made from green products (trees). Yellow is the color chosen for glass products and its association with sunshine and its natural sourced silica from sand.

Natural Additives, Pigments and Dyes

There are also greener colorants that are used in a variety of industries. Plant and animal colorants can be traced back as early as 10,000 BC. However, in the pastdecade, there has been a renewed interest in using natural dyes in the fashion industry since consumers are concerned about the toxicity of their synthetic counterparts. Most natural dyes are derived from plants and fungi. There are many dyes that come directly from fruits and vegetables. Wood and bark are additional sources of plant dyes. Different types of textiles require different types of dyes and processes. For plant-based textiles, such as cotton, hemp, bamboo and rayon (synthetic silk), dyes must be fixed with oxygen, sunlight or salt. For animal-based textiles such as silk, wool and leather, dyes require a vehicle that acts as a bonding agent. When designing colors for a plastics application, the prospect of using a natural dye or pigment is quite challenging. Crystalline polymers such as polypropylene and polyethylene cannot be colored with dyes because they cannot retain the dyes within their matrix and the dyes will migrate or bloom to the surface of the plastic. Dyes can be used in amorphous polymers such as polystyrene or polycarbonate but many of the dyes are not stable at the higher temperatures required to process these types of polymers and will break down as the polymer is in the melt phase.

Further Reading

http://www.chinabakelite.com/en/index.asp
http://www.wisegeek.com/what-is-bakelite.htm#didyouknowout
http://www.who.int/csr/media/faq/QAmelamine/en/
https://en.wikipedia.org/wiki/Melamine
http://recyclingrainbow.com/

43 Regulatory Restrictions

Regulatory restrictions are established to protect the health, safety and wellbeing of consumers and manufacturers. These regulations become tighter every year and different parts of the world have more stringent regulations than others. Also, different industries have tighter regulations or different regulations than others. This section will detail some of the history and specific types or regulations that play a role in protecting the products of those industries.

Food

There is a long history of color additives for food and most of these were natural products or extracts of natural products. There are many naturally pigmented foods that are brightly colored that were used as extracts to color other foods. Pomegranates, spinach, berries, beets and carrots were quite common and are still in use today.

However, natural does not equal nontoxic. Early food colorants consisted mostly of plant, vegetable and fruit colorants but sometimes metallics were used as garnishes or even coatings for the upper classes of society. As an example, leafed golds and silvers were applied to foods and these are nontoxic but if adulterated versions were used they could contain mercury or tin oxide.

Smalt is a cobalt potassium borosilicate glass that is typically used to pigment blue artist's paints and is the primary component of cobalt glass but was also used to color food. Yellow smalt that was also a common food colorant is a sodium or potassium cobaltnitrite that is also toxic. Although minute traces of Cobalt are essential for health (vitamin B12), larger amounts of Cobalt can be highly toxic and its ingestion/exposure has been attributed to deaths from cardiomyopathy and contact dermatitis.

"Naples yellow" which is a lead tin antimonite pigment was also used as a garnish for foods. Lead and antimony are both quite toxic even in small quantities.

Color Trends and Selection for Product Design.
DOI: http://dx.doi.org/10.1016/B978-0-323-39395-9.00043-8

Gamboge is a yellow gum resin that is tapped from trees that are native to Cambodia. In addition to an early food colorant, it is still used to color textiles and is commonly associated with the saffron or mustard color of Buddhist monk robes. Brindleberry is the fruit of these trees and is frequently used to cure fish and make sour curries. In high concentrations, Gamboge can be toxic, causing liver damage. In smaller concentrations, the yellow extract can be used as an emetic and was touted as a weight-loss agent according to Mehmet Oz (Dr. Oz).

Daylilies were also used to color food and in some cultures are still used as colorants and as foodstuffs. Chinese cultures dry the flowers and sell them as Gum Jum that is used for daylily soup and some common Northern Chinese dishes such as moo shoo pork. Not all lilies are toxic but some can cause vomiting, diarrhea, seizures, liver failure and death.

Lapis and azurite were also used as a blue food colorants. In some cases, Lapis lazuli stones were ground into a fine paste with oil and were not toxic but azurite (blue) and malachite (green) are hydrous copper carbonates. The high copper content of both azurite and malachite makes them a mild toxin in some regulatory spheres. Ingestion and exposure to copper has attributed to illnesses ranging from dementia to liver failure.

Cinnabar, mercury sulfide, is a beautiful red pigment that was also used to color foods. Synthetic Cinnabar (Pigment Red 106) is still used in China as a colorant in paints and varnishes but is banned in most parts of the world due to its toxicity. Mercury sulfide colorants were replaced by Cadmium Red in the 20th century. Red and Orange Cadmiums were strong and clean shades and were much more lightfast than Cinnabar but were also banned in automotive paints due to their inherent toxicity when inhaled or ingested. There are several different types of cancer that have been traced to cadmium exposure or ingestion as well as liver and kidney failure. Cadmiums were replaced by Azo colorants for industrial applications which are also quite light stable when compared to Cinnabar but lacks the brilliant coloration of its predecessors.

Toys

The toy industry is highly regulated and has some of the most stringent guidelines for manufacturers. Most toy designers are familiar with the choke tube. This is a small plastic tube that is used as a guide for size of the product. If the toy or a piece of a toy fits inside the choke tube then it is considered to be a choking hazard and the packaging must clearly state that the toy is a choking hazard. Marbles, small balls, plastic building bricks and puzzle pieces frequently fall into this category. Small magnets are also a problem since they frequently contain rare earth metals that are classified as toxic if they are swallowed or if multiple magnets are swallowed they might join to become a larger object that can cause intestinal blockage or damage. Balloons and plastic films on the surface of toys can be a choking hazard. Strings and straps can be a strangulation hazard. Colors used in and on toys are subjected to the same stringent guidelines because children frequently put toys in their mouths or ingest small parts of the toys that are either bitten of or break as a result of play. Although most recalls of toys are related to choking or ingestion hazards, quite a few products still are imported with paints containing lead or lead being used a source of metal in children's jewelry and craft kits. Chromated Copper Arsenic treated wood are frequently used in large outdoor play structures and decks and are still widely used.

Further Reading

http://www.witchipedia.com/mineral:azurite
https://en.wikipedia.org/wiki/Cadmium_poisoning
http://www.slate.com/articles/health_and_science/medical_examiner/2013/01/can_you_trust_dr_oz_his_medical_advice_often_conflicts_with_the_best_science.single.html
https://en.wikipedia.org/wiki/Cobalt
http://www.pbm.com/~lindahl/articles/food_coloring_agents.html
http://dictionary.reference.com/browse/gamboge
https://en.wikipedia.org/wiki/Garcinia_gummi-gutta
http://www.naturalpigments.com/lead-white.html
http://www.controverscial.com/Malachite.htm
https://en.wikipedia.org/wiki/Natural_dye
https://en.wikipedia.org/wiki/Potassium_cobaltinitrite
https://mistralhowto.wordpress.com/tag/pigment-red-105/
http://www.toyassociation.org/TIA/Priorities_Policy/safety/Priorities___Policy/Industry_Priorities/Product_Safety/Standards.aspx#.VosXI6Eo5gV
http://ec.europa.eu/growth/sectors/toys/safety/index_en.htm

44 Evolution of Pigments and Dyes

As mentioned previously, the earliest use of colorants was usually derived from, plants, fruits and vegetables. There were also some animal products derived from bones, blood, hair and fecal material. Many minerals were also used in cosmetics. It is beyond the scope of this book to outline the entire evolution of colorants throughout the ages, so a snapshot of the 20th and 21st centuries of some of the most important colors will be discussed.

Heavy Metals

Heavy metals are better defined for the context of this discussion as toxic metals. Technically, heavy metals are defined as any metal having a specific gravity greater than 5.0. However, not all heavy metals are toxic. Gold and silver are both classified as heavy metals but are by no means toxic. Some of the most common toxic metals are lead, mercury, cadmium, arsenic and chromium. Other heavier metals that have noted levels of toxicity are zinc, iron and copper. It is important to note that many of these metals are already in the human body in trace amounts and are essential for survival. However, when humans are exposed to higher levels of these metals, adverse reactions can occur. Many pigments have metals as part of chemical structure and some heavy metals, such as lead, were used as pigments in cosmetics and other products before their toxicity was well-understood.

Lead

Lead White, also known as lead carbonate is a naturally occurring mineral called *cerussite*. It is a beautiful white pigment that has been used for hundreds of years in paintings, cosmetics and industrial paints. Lead carbonate is still commercially available and widely used in the casting process to create buttons for clothing. Lead Oxide Red (Red 105) is a red or orange-colored pigment that is used as a primer in paints and a rust

Color Trends and Selection for Product Design.
DOI: http://dx.doi.org/10.1016/B978-0-323-39395-9.00044-X

inhibitor. Due to their high-heat stability, they are frequently used as pottery glazes. These lead products are highly toxic and should not be ingested or inhaled. The Lead-Free Toy Act was incorporated into the larger Consumer Product Safety Improvement Act and was signed into law in 2008 in the United States to protect citizens from the use of toxic substances, such as leads and phthalates, in many common types of products including toys, school supplies, jewelry, shoes, etc. The production of lead-containing products is still wide-spread and very difficult to test and control globally.

Chromates

The chromium compounds were popular, in part, because of their ability to shift color when combined with other compounds. Crocoite is a naturally occurring mineral of lead chromate that yields a bright orange color. In the 1700s, it replaced *Realgar*, an orange, very toxic arsenic-based compound. Chrome Yellow is made by replacing the lead component with zinc or strontium. When heated, Chrome Yellow shifts to a bright green.

The toxicity of chromium metal depends on its oxidation state. Chrome III (chromite) is not considered toxic at low levels and is even approved as a vitamin supplement. However, Chrome IV, V and VI are considered toxic Chrome Yellow, Chrome Orange and Chrome Green are three colorants that contain chromates. There are many other colorants that also contain chromates such as Molybdate Orange (Red 104), Barium Yellow (Yellow 31) and Strontium Yellow (Yellow 32). All of these colorants also have additional toxicity due to the presence of lead. Chrome Yellow is a high-hiding bright yellow pigment with moderate lightfastness that is cateogorized as a Lead III Chromate. It is a color frequently associates with school buses and taxi cabs. Due to its inherent toxicity, it was replaced by molybdenum compounds that were later replaced with Azo colorants when the molybdenum compounds were classified as carcinogens. Chrome Green (Pigment Green 17) is also known as chrome oxide green and is a Chrome III compound and is not considered toxic by current standards. Chrome Orange is also known as Chrome Red and is a compound that consists of lead carbonate and lead oxide. The chromium is in the form of chrome VI (hexavalent) so it is very toxic due to its lead and chromate components.

Molybdates

Orange and red shade molybdate pigments were frequently used in plastics paints and coatings throughout the 20th century due to their intense shades, high-hiding and superior lightfastness. These colorants were extremely popular automotive paint shades in the 1960s and 1970s for muscle and pony cars. These products are created by combining lead chromate and lead molybdate and their use is highly restricted due to their lead and chromate components. The red shades contain a higher ratio of lead molybdate and the yellow-orange shades contain a higher ratio of lead chromate.

Antimony

Antimony Yellow or Naples Yellow was one of the first synthetic colorants ever produced in the 1600s. It is composed of Lead II Antimonate and has dull, earth-tone yellowish appearance. Antimony Sulfide was used as a cosmetic product (Kohl) by the early Egyptians. Antimony Tin Oxide is used as a laermarking pigment for plastics. Antimony is classified as a toxic substance in the United States and is known to affect the respiratory and cardiovascular systems.

Types of Regulations Related to Colorants

Regulatory requirements for colorants are very broad and vary by geography, market, product and culture/religion. There are also local regulations about color, specifically zoning laws, requiring the use of certain colors and the banning of others. As an example, there are certain towns in Germany that require all home owners to have a red roof on their homes. In the United States, colorants are either certified or exempt (from certification). These certifications are published in the Code of Federal Regulations (CFR) issued by the Food & Drug Administration (FDA). Natural colorants (those inherent in the food or cosmetic product) are generally exempt from certification. Certified colorants are those that have been tested and deemed safe for certain types of applications. Natural colorants that are added to a product to enhance the color would require certification. As an example, if a dye extracted from pomegranates was used to boost the colors of a fruit punch then that dye would require testing and certification by the FDA.

WEES Directive

The Waste Electrical and Electronic Equipment Directive was created in the EU in 2012. This directive was established to address the increasing amount of electrical and electronic waste that has infiltrated municipal waste streams in the past few decades. Two major problems with this type of waste are that the products are semidisposable and have life cycles that range from 6 months to 3 years and that many of them contain heavy metals and trace amounts of toxic substances. This directive was established to help consumers dispose of these products correctly as well as to help manufacturers create reusable, recyclable and greener products. There are 10 different categories of products that are encompassed within this directive including, computers, telephones, toys, sporting goods, medical devices, household appliances, lighting, etc. Although the emphasis is currently on the electronic components of these products, it could eventually extend to the colorants used in and on these products as well.

REACh

In 2007, the European enacted REACh. REACh is an acronym for *R*egulation, *E*valuation, *A*uthorization and restriction of *Ch*emicals. This regulation is designed to protect European citizens from hazardous substances and products in the marketplace. This regulation is not only important for those companies that manufacture products in Europe but also for those companies that manufacture in other parts of the world and export products into the EU. When designing new products, it is best to specify raw materials that are REACh compliant to avoid regulatory issues if these products enter into the EU market.

ROHS

In 2003, the European Union enacted the *R*estriction *O*f *H*azardous *S*ubstances for electronic components and products. This initiative was enacted to restrict, reduce and monitor the pollution caused by electronic products. The list of hazardous substances is quite small and includes Lead, Cadmium, Chromium (VI), Mercury and certain phenyl and phthalate compounds.

China ROHS (ACPEIP)

In 2006, China created their version of the European ROHS legislation. In some instances, this legislation is more restrictive that the European initiative since it requires labels on EIPs (electronic information products) and additional hazard information that are made in China. They are also establishing more stringent testing methods that are currently required in the European version of the original ROHS initiative.

CONEG

CONEG is the *C*oaltion *Of* *N*orth *E*ast *G*overnors in the United States. This group of seven Governors from the northeastern states of Maine, New Hampshire, Vermont, Massachussets, Connecticut, Rhode Island and New York introduced a regulation for the restriction of heavy metals in their states for packaging products. CONEG introduced Toxics in Packaging Clearing House (TPCH) in 1992 to help other states that wanted to adopt the CONEG regulation. There are 19 states in the United States that have adopted this regulation to date. The regulation phases out the use of heavy metals in packaging including lead, cadmium, mercury and chrome VI.

Kyoto Protocol

The Kyoto Protocol is an international directive that was enacted in 2005 as an extension of the United Nations Framework Convention on Climate Control (UNFCCC) that was developed in 1992. These initiatives were established to reduce the emissions of greenhouse gases. Specifically, the Kyoto Protocol targets methane, nitrous oxide, sulfur hexafluoride, Hydrofluorocarbons (used mostly as refrigerants), Perfluorinated compounds (Teflon, Scotchguard and other stain and stick-resistant compounds) and Carbon Dioxide.

End of Life Vehicle Directive

In 1997, the European Union drafted the ELV Directive to reduce waste from vehicles that are end of their usefulness. The primary directives of this legislation addresses the restrictive use of toxic materials including, but not limited to, chromates, cadmium, lead and mercury. This includes colorants that are used on or in automotive parts and finishes. Another tenet of this directive includes the proper disposal of vehicles and vehicle parts to insure they are recycled correctly. All automotive fluids and materials that are potentially hazardous are also neutralized prior to disposal.

Conflict Minerals Policy

In response to violence and human rights violations in the mining of certain *minerals* from the "Conflict Region," which is a geographical region that uses profit generated from sales of natural materials to perpetuate war and violence. Some of the more common raw materials are gold and diamonds but there are many other minerals and rare earth elements that are used in electronics, magnets, batteries and pigments that are also impacted. In 2010, the United States passed the Dodd-Frank Wallstreet Reform Act which required global corporations to investigate their supply chain of raw materials and report on their use of conflict minerals in their manufacturing and sales processes.

Animal Derivative/Bovine Derivative

Since the "Mad Cow" or BSE (Bovine Spongiform Encephalopothy) crisis of the 1990s, more attention has been paid to animal content in foods, cosmetics, pharmaceuticals, packaging and consumer products. Animal products are commonly used in the manufacturing and processing of plastics and plastics packaging. Most of these are in the form of stearates that are long chain fatty acids which are added during the manufacturing of polymers to impart lubricity. These stearates are derived from beef fat and tallow and are laked to various salts such as calcium, magnesium or even metallized salts such as aluminum. These slip agents are also used during the molding process to help plastic products release from the walls of the mold in the injection molder. Since these stearates are generally regarded as a waste product from the meat industry, they are quite inexpensive and are commonly used. Nonanimal-derived stearates are also available but are considerably more expensive than their animal counterparts. Stearates are also sometimes used as stabilizers for color pigments to prevent plate-out with flexible PVC as well as prevent degradation of colorants at higher processing temperatures. It is important for designers, engineers and marketers to understand that if they wish to make claims of their products as being vegan or animal-free, they need to investigate the components of their raw materials before asserting these claims.

EU Toy Safety Directive 2009/48/EC (New Toy Safety Directive)

This directive outlines the specifications for all toys that are marketed in Europe regardless of their country of origin. There are both general and specific risks that must be observed. General risks consist of the health and welfare of the children who play with the toys as well as the caregivers and parents and other children who come in contact with the toys. The specific risks consist of safety covering broad categories of chemical and mechanical, flammable and electrical, and physical and radioactivity. Chemical restrictions are further specified as carcinogens, mutagens and in utero hazards. This also includes 19 heavy metal additives that are sometimes used in colorants. Allergens and certain fragrances are also restricted.

CPSIA and RTP Toy Standard

The Consumer Product Safety Improvement Act of 2008 (CPSIA) was enacted to provide safety standards for children's products in the United States. There are currently over 100 tests and restrictions for children's toys that are sold in the United States that are either manufactured domestically or anywhere else in the world. Some of the generic tests are for choking. From a color perspective, the most important component of these is the acts banning the use of lead in paints and plastics and formaldehyde in textiles and apparel. Toy legislation in the United States is the most stringent and developed than any other country in the world and as a result many other countries have adopted this slate of legislation for their own consumers. RTP is a subsection of the CPSIA and concerns the reasonable testing procedure for the following areas: periodic testing of existing products, material changes, undue influence and proper record keeping of testing procedures and results.

KFDA

The Korean FDA (KFDA) does not regulate pharmaceutical products but does regulate all cosmetics and related products as well as soaps, ox bile, gels and capsules used in the production of pharmaceuticals for import or manufacture in Korea.

EU Food Contact Compliance

EFSA (European Food Safety Association) is responsible for monitoring and issuing compliance certificates for food products and their associated packaging. Their aim is to reduce or eliminate and contaminants or harmful chemicals that come into contact either directly or through migration that will impact the flavor, texture of the food or impair the health of consumers. This includes colorants and flavors and ranges from simple packaging to coffee brewers to food processing equipment.

Kosher

Certain types of products (usually food, cosmetics and personal care items) may require the use of Kosher certification and should be taken into consideration prior to production. Colorants that contain animal products such as cochineal that are derived from insects are forbidden. Also colorants that are spray-dried in equipment and that are not certified by a rabbi are also not Kosher since those spray dryers are frequently used to dry other non-Kosher colorants. Also, colors that contain plant products associated with leavening products (like yeast) are not certified Kosher for the Passover holiday. Caramel colors are also sometimes not Kosher. It is best to inquire from the colorant manufacturer if their colorants are certified Kosher before using.

Halal

Halal is the Islamic regulation that refers mostly to food but also relates to cosmetics, pharmaceuticals and other personal care products. Halal literally means permissible in Islam. Colorants also fall under Halal guidelines and should not contain animal products or derived from alcohols. As an example, Tumeric is considered to be questionable since it might be made with animal fat-based emulsifiers. Cochineal is another colorant that is not Halal since it made from insects.

Montreal Protocol

The Montreal Protocol is an international treaty that was adopted in 1987 after the discovery of a "hole" in the ozone shield over Antarctica by the British Antarctic Survey in 1985. The protocol was initially signed by approximately 25 countries but is currently held by more than 200 countries. The protocol required that members reduce their use of halogens, chlorofluorocarbons and other synthetically derived ozone depleting chemicals. In 1989, the protocol was expanded to include the reduction and the eventual ban of the production of carbon tetrachloride and methylene chloride.

FDA

The Food & Drug Administration is a government agency in the United States that is responsible for monitoring and approval of all raw materials in food, drugs, cosmetics, medical devices and their associated packaging. There are different levels of certification from direct use in a product to indirect contact with a product. The CFR outlines which colorants can be used for what types of products and also their level of usage. It is best to use FDA-approved colorants whenever possible since their usage limits and certifications are recognized globally.

California Prop 65

The safe drinking act and environmental toxins act are commonly known as Proposal 65 and were passed in California in 1986. This proposal requires the State of California to publish a list of chemicals known to cause cancer, reproductive harm or birth defects. There are currently over 800 chemicals on the list including some colorants. The list is published by the Office of Environmental Health Hazard Assessment (OEHHA) and part of California's Environmental Protection Agency.

TSCA Certification

The Toxic Substance Control Act (TSCA) is administered by the Environmental Protection Agency in the United States. The TSCA regulation was enacted in 1976 and oversees the use and regulation of colors and chemicals in different products and as raw materials. TSCA was designed to ensure that products are safe for intended use as specified by the manufacturer and to preserve the American consumer's confidence in products that are currently in the marketplace.

Health Canada Submission

Health Canada is part of the Canadian government that strives to protect and educate Canadian citizens about their health and welfare. There are 12 different branches of Health Canada and two of them that regulate products and services are the Healthy Environments and Consumer Safety Branch (HECSB) and the Health Products and Food Branch (HPFB). The HECSB serves to educate Canadian citizens about healthy environments and products and regulate products and services that impact the health and welfare of Canadians. The HPFB endeavors to monitor the risks and benefits of health and healthy food products. Canada Health is similar to the Food and Drug Administration in the United States in that they seek to mitigate risk factors and eliminate unsafe consumer products, packaging and food.

NSF

The National Sanitation Foundation (NSF) is an international organization that started in 1944 to protect consumers against unsanitary conditions in soda fountains and luncheonettes in the United States. It is now a global protective agency that provides safety standards and guidelines for many different industries including drinking water, pharma and cosmetics.

Further Reading

https://en.wikipedia.org/wiki/Restriction_of_Hazardous_Substances_Directive
http://aercmn.com/blog/posts/understanding-the-toxicity-of-lilies/
http://www.prop65clearinghouse.com/?page=prop_65_plain_english&IdPage=0&IdCase=0&IdLink=0&IdDefendant=0
http://www.canadianhalalfoods.com/e-numbers_list.html
http://www.cyclopaedia.nl/wiki/Chumetz kosher
http://www.nsf.org/about-nsf/mission-values-history/
http://ec.europa.eu/growth/sectors/chemicals/reach/index_en.htm
http://www.rohsguide.com/
https://oukosher.org/blog/consumer-kosher/the-kosher-color-spectrum-the-kosher-status-of-food-colorants/
http://www.naturalpigments.com/orange-molybdate.html
https://en.wikipedia.org/wiki/Waste_Electrical_and_Electronic_Equipment_Directive
https://en.wikipedia.org/wiki/Conflict_resource
http://www.efsa.europa.eu/
http://www.cpsc.gov//PageFiles/129663/cpsia.pdf
http://www.doverchem.com/Products/MetallicStearates.aspx
http://www.britannica.com/event/Montreal-Protocol
http://www.ghgonline.org/othershfcs.htm (green house gases)
http://en.tdk.eu/tdk-en/180308/company/environmental-protection/china-rohs
https://en.wikipedia.org/wiki/Lemon_law

45 Applications of Color

Functional Uses of Color

Frequently, colorants are used for other purposes beyond an esthetic effect to a product. A prime example would be the use of dyes to color lenses of tail lights in cars so that it alerts on-coming traffic to a stopped or slowing vehicle. Another example would be where spices such as turmeric, saffron and cinnamon are used to enhance the color of food but also impart unique and subtle flavors to the prepared foods. Colorants are also sometimes used as tracers in products to prove authenticity of a brand or to test for the presence or concentration of a certain ingredient.

Coil Coating

Coil coating is a method of prepainting metals prior to installation for steel and aluminum metal sheeting. After the metal is processed and rolled, the rolls are unrolled, primed, painted with liquid paint and rerolled. These rolls may also be texturized or embossed to create patterns on the surface of the sheets. Coil coating is considered to be a green method of coating since prepainting eliminates costly and environmentally damaging on-site painting. It is also attractive to architects and construction companies since it is a LEED (Leadership in Environmental & Energy Design) certified green material. This metal sheeting and panels are frequently used in roofing and siding architectural applications.

Powder Coating

Powder coating is a method of coating that resides somewhere between paints and plastics. This method involves electrostatically applying colored plastic powders onto a charged surface (usually metal). Initially, the polymer (polyester, epoxy, etc.) is extruded with colorants and additives and then pulverized into a fine powder. This is high-shear

Color Trends and Selection for Product Design.
DOI: http://dx.doi.org/10.1016/B978-0-323-39395-9.00045-1

process so pigments such as pearlescents will need to be blended to the powder prior to coating to prevent damage to the luster effect. The dry powder is pushed through a hose and into a spray nozzle where it is electrostatically charged, allowing the powder to temporarily adhere to the metal part which is also charged to attract the powder. Once the metal part is covered, it is placed in a large oven for baking. This process can take several hours to melt the powder and permanently bond to the surface of the part. This creates an incredibly durable coating that is highly resistant to chips, cracks and scratches. Powder coating is frequently used on appliances as well as on the bottom exterior panels of vehicles that are subjected to a great deal of damage from the road. Ideally, powder coating would be a good candidate to replace all exterior automotive paint but this process does not lend itself to a smooth and glossy Class A surface finish which is currently in trend for automotive specifications. Powder coating is sometimes touted as a greener painting method than traditional liquid paints due to its lack of volatile organic components that become airborne during application; however, powder coating does create a great deal of dust while applying which is an inhalation hazard as well as an air contaminant. Additionally, when applying the powder, a large percentage of it falls off and is collected in a waste bin. Depending on the components of the formulation, sometimes it can be reused and sometimes it must be discarded since reuse can cause a nonhomogenous dispersion of additives. This can result in a large amount of spray to waste that is costly and certainly not green when discarded. Also, when baking these epoxies and polyester powders, volatiles are evolved and can cause additional environmental concerns.

Injection Molding (Molded in Color)

Injection molding is one of the most common methods of creating plastic parts. The method is quite simple. Plastic pellets that contain all additives and colors are poured into a hopper or bin at the back of the machine. A measured amount of pellets are released into a heated chamber that contains a large spinning screw. The plastic pellets are mixed and melted as they are conveyed along the flights of the screw and become a molten plastic by the time they reach the tip of the screw. The end of the screw has a pointed section that pushes the molten plastic into the closed mold that bears a hollow cavity in the shape of the plastic part. A small amount (dose) of molten plastic is injected into the mold and remains until the plastic cools and solidifies.

The two sides of the mold then open and the plastic piece is pushed out of the mold with little ejector pins. The plastic piece is usually collected underneath the machine or in some cases falls onto a conveyor belt and sent to another part of the production process for labeling or filling.

Injection molding is the method of choice for mass-produced pieces where thousands of plastic pieces can be produced daily. The startup costs of injection molding are quite high. The machine itself is quite costly and requires a great deal of expertise to operate and maintain. Every plastic piece requires its own unique mold. These molds are generally fabricated from hardened steel or other durable metals. These molds are also quite expensive and are custom made to fit into a specific injection molding machine or type of molder. The surface finish of these molds is critical to determine how the colorants will appear in the molded piece. Generally speaking, smoother finishes will create the best appearance when using effect pigments. Whenever possible, it is best to determine the colorways of the final products prior to mold design so that the best effects will be achieved. A chromed mold or highly polished steel mold will yield the best effects since many pearlescent and metallic pigments are platey and will align flatter on the surface and create a high luster finish of the molded part rather than disorient or scatter on the surface when the mold has a textured, matte or unpolished mold surface. These unpolished mold surfaces create a less-lustrous effect in the final molded part. Also, when textured patterns are added to the surface of the mold, care must be exercised to apply the texture evenly and not overlap since any error in creating the texture can result in a phenomenon called tiger striping on the surface of the finished part. Tiger stripes are subtle bands or stripe that can appear lighter or darker on the molded piece. Some of this can be corrected by changing the molding conditions but generally speaking the best method to eliminate this flaw is by resurfacing the mold. Tiger striping can also be a result of a poorly vented mold where air becomes trapped between the melt and mold surface, creating a rippled surface where the molten plastic cannot properly adhere to the surface of the mold. One of the biggest problems with injection molding is flow lines. Flow lines occur when the melted polymer flows into the mold and platey pigments or poorly dispersed colorants align on edge and create unsightly lines in certain parts on the surface of the molded part. There is an entire field of study around eliminating or reducing flow lines but it is beyond the scope of this discussion. In brief, the best methods to eliminate flow lines are to design the mold correctly prior to molding the plastic parts or to reduce or eliminate the use of small platey effect pigments in the initial color design.

Multilayer Coatings and Plastics (Candy Coats)

Multilayer paints and plastics are used when high quality, multifunctional and greener requirements must be met for specific applications. In consumer packaging, multilayer blow molded bottles, tubes and pouches are common. For bottles and tubes, multilayer technologies can offer cost-savings by adding colorants in one or two layers of the product rather than coloring the entire product. Additionally, one of the middle layers can contain reground or recycled polymers that can reduce costs and create a greener product (regrind). One of the greatest advantages of these multilayer products is increased esthetics. Designers can put absorption colors (reds, blues, greens, etc.) in an inner layer, behind a layer that just contains effect pigments such as semitransparent micas or color shifting products and then put a final layer of a clear coat as an outer skin to achieve a glossy topcoat, similar to an automotive paint finish. Another high-end coloration technique is called candy coats where a metal flake (aluminum, brass, etc.) is added in the inner layer, topped with a highly transparent dye or colorant and then an additional clear coat on the surface. This configuration allows the metallic layer to reflect through the colorant and create a vibrant high-tech effect.

Airbrushing (Cosmetics)

Airbrushing is simply spray painting where a small gun (usually in the shape of a pen) is used to apply thin coats of products to a surface with great detail using compressed air that is attached to the gun with a small hose. Traditionally, airbrushes were developed for brushing watercolors onto the surfaces of art canvases but later their use and applications blossomed from illustrations to photographic manipulations and custom applications for automotive and hobby painting. Although airbrushes have been used for cosmetic application since the 1950s, it has become essential for applying cosmetics since the advent of high-definition TV where every flaw and facial defect becomes magnified on the larger screens. Vibrant and theatrical coloration is easily applied for special effect makeup, clowns, temporary tattoos and spray-on tans.

Further Reading

http://www.coilcoating.org/index.php/education/tutorials#prettyPhoto
http://blog.traininteractive.com/plastics_training_technology_information/2013/11/correcting-tiger-striping.html
http://leedcert.com/
https://en.wikipedia.org/wiki/Powder_coating
http://www.urekem-paints.com/candy-paint-colors

SECTION IV

Color Selection Process

46 Color Selection by Market

The color selection process varies with different industries and each industry has different timelines for their colors. This section will look at product colors for different industries and their markets, how different industries select colors and use color trends and how color selections change over time.

Fashion

As was stated earlier, fashion colors tend to be a the fore-front of trends with couture fashions being well ahead of ready to wear and followed by mass-market clothing. Couture clothing has rapid color cycling and changes dramatically with the seasons, geography and more recently political events. Fashion is also influenced by music culture and religion. Recently, there has been a trend for high-end fashions to become more modest. This includes a greater use of scarfs, head and body wraps as well as pleats and ruffles. This could be due, in part, to the modesty requirements of the Islamic religions and the affluent societies that they represent. In addition to the modesty element of these styles, we are also colors that are representative of those cultures, earthtones, sandstones, exotic spices, etc. Clearly, there is nothing new about spice colors, modest clothing or Islamic cultures but having them coalesce in a high-end fashion movement is certainly indicative of a trend. Another example is when technology influences fashion and color choices. At the S/S 2016 RTW in Paris, Kunihiko Morinaga, lead designer from Anrealage, showcased a series of garments that were to be viewed with the flashlights on the audience member's smartphones that enhanced the reflective nature of the garments. Initially, the fashions appeared black and quite dark but lit up and textures and patterns emerged when enhanced by the additional light sources. The inspiration was that reality is reflected through the fashions and that this reflection brings dreams into reality. Currently, fashion designers also use a variety of other inspirational sources such a trend consultant agencies that scour the art and tech worlds for new sources of ideas and present these colors,

Color Trends and Selection for Product Design.
DOI: http://dx.doi.org/10.1016/B978-0-323-39395-9.00046-3

textures and materials as pallets to the fashion houses, usually as mood or inspiration boards for specific types of fashion. Couture designers also do a great deal of on-line research to study their competitors and other industries as well as attend conferences, tradeshows and entertainment events to create their pallets. In the early days of couture, fashion colors were often dictated by the affluent consumers who reflected the societal trends of their day. In modern times, these consumers (and many other industries) look to the fashion houses for the latest and most progressive styles and color trends. Modern couture designers also look to the runway shows of design schools such as Parson's School of Fashion Design or Fashion Institute of Technology in New York as well as Central Saint Martens and Royal College of Art in London.

Ready to wear (RTW) fashions sometimes take their color cues from couture but also tend to favor more classic colors and styles. Colors tend to be darkened more subdued shades from the couture lines since these consumers tend to wear the clothing for more than one season. Also, better menswear and business attire tend to fall into this category so those fashions need to be more conservative in style and coloration.

Mass-market clothing tends to take color and fashion trends to the extreme where they will "borrow" trend information from couture and exaggerate the colors and styles. Mass-market children and teenswear are good examples of this over-amplification of style and color trend. Another side of mass-market manufactures utility clothing, uniforms, workwear that is almost entirely devoid of color and fashion trend.

Automotive

The automotive design world generally seems to operate in two different spheres. The exterior finish group is very independent of the interior design group in most instances. As mentioned previously, there are currently very few options for color pallets for exterior automotive of mass-produced vehicles. However, these designers are constantly looking at special effect colorants to modify their pallets of black, white and gray/silver. Generally speaking, this is a very cost-prohibitive market so most of the colors or colorways that are introduced on concepts cars never make it into actual production. Cost is only one reason this occurs. This distillation of color might also be in part due to the instability of the colorants or the difficulty or expense of applying the exotic colorants in a production environment. Every few years, a new color will emerge in the pallet due to a new technology or chemistry that creates a unique color or effect. As an example, several manufacturers have developed effect pigments that fall into the dark orange or burnished copper color space in recent years. Exterior automotive designers would like to offer bright oranges that were once prevalent on sports cars a few years ago but were pulled out of production due to the heavy metal contents of these colorants. These newer effect pigments are less of a health and environmental hazard but much more expensive than their predecessors. They are also not as stable or bright and clean in shade.

The interior automotive designers are given much more freedom to experiment with new colors and materials since the consumer (and the dealers) tends to choose the car colors based on the exterior. This allows them to be less risk-averse than their exterior counterparts. Interior designers are very concerned with color harmony and ensuring that one color in the interior closely matches or complements the other colors in the cabin. This is a massive undertaking since many different pieces are manufactured by many different companies all over the world. There are also differences in texture and gloss and durability that must be taken into account to ensure a successful design project.

Both interior and exterior designers look to their suppliers or tradeshows for new colors and materials. They also might use an outside design house for sources of inspiration. Typically, a design group will work with their corporate marketing group to determine a strategy for a new design for a new vehicle or a refresh for an existing vehicle. Sometimes, the marketing group will dictate a theme, demographic and even a color and material pallet for the designers to work with. Other times, they might offer some conceptual themes for the designers to follow such as "soccer moms," "mid-life crisis" or "urban cowboy." The designers may take these themes and work them into mood boards with a variety of colors, textures, words, images, etc. as a starting point. Once developed, the designers might work on pulling master standards of old colors and materials or might look for something entirely new and call in their suppliers for a design brief where they will review their strategy and ask the suppliers to submit proposals and samples. This process is very similar to those used in other industries as well. However, the automotive timeline can take years to complete since the automotive industry requires extensive safety, regulatory and stability testing on all of its components. In some instances of smaller and more agile manufacturers, the designs and color selections of the interior and exterior part of the vehicles might be handled by one design group. Once the designers select their products and colors, the project is given to the engineers to test the materials and run trials. If the materials and colors fail, the designers will need to go back and choose more suitable options. This is a frequent push and pull between the designers and engineers where the designers are looking for new and innovative products to enhance their designs and the engineers tend to favor the tried and true materials and colors. Unfortunately, this leads to newly designed products that do not appear or function much differently than their predecessors or competitors products.

Cosmetics

The cosmetics market color selection is closely dictated by fashion and consumer complexion. Like other industries, they have in-house design teams that develop and implement strategies for new products and refreshed products. They also rely on outside consultants and suppliers to offer up new ideas for colors and technologies. Since fashion models need to be properly groomed on the runway, it is essential that their makeup colors complement the new fashions so these cosmetic designers pay very close attention to the colors on the runways and the color trends that are emerging for fashion.

Housewares

Housewares might be loosely defined as tools and objects used in the home that are not furniture or appliances. Retailers generally break these down into soft goods like textiles, throw rugs, bedding, curtains, wall coverings and hard goods; kitchenware, cleaning supplies, storage and organizational racks and bins. Sometimes after-market automotive accessories fall into this category as well. Houseware colors and trends are heavily influenced by retailers who handle their products. Two good of examples of this are Target and Ikea.

Although Ikea is known to be a Danish furniture manufacturer, their retail stores sell well beyond furniture and include many soft goods, hard goods, food, clothing and toys. Ikea hires many textile and furniture designers. Most of their manufacturing is outsourced to smaller designers who make the products for Ikea. Ikea also employs their own color experts who predict colors for their products several years in advance. They also rely on suppliers and consultants to bring in the latest color and technology trends.

Target is somewhat different. They also have many designers that create new textiles and furnishings for their stores but Target is completely outsourced. They do not directly manufacture any of the products they sell in their stores. Target has an interesting history of collaborating with celebrity designers such as Michael Graves (an American architect) for housewares and Missoni an Italian luxury fashion brand that created designs for mass-market clothing, houseware and tableware. These types of celebrity collaborations are good for both sides as high-end fashion continues to lose market share and publically held retailers like Target need to constantly reinvent themselves to increase profits and brand-recognition. Although the concept of bringing high-end style to the masses is a good one, most Target customers barely get a glimpse of these high-fashion remakes since their production is time-limited and relatively small runs. Most of these products are bought up in bulk and re-sold with massive mark-ups on auction sites. However, the collaborative concept is still a good one for retailers and designers.

Electronics

The electronics market colors tend to mirror, automotive and menswear colors. It is similar to a three-way mirror where each industry looks to the other for inspiration and collaboration. They also collaborate with jewelry designers and brands. As an example, Vertu is a cellphone manufacturer (Android platform) that has formed a 5 year collaboration with Bentley automotive. These phones are mostly about style, covered in diamonds and exotic animal skins but also offer some interesting functional components such as their scratch-proof screens made of sapphire crystal similar technology is used in high-end watches. Blackberry is another company that is offering some luxury smartphones in their collaboration with Porsche. Again, the collaboration is more about style than substance but Blackberry still remains in the market, mostly due to their nonvirtual qwerty keyboard.

It is rumored that Apple will be launching a pink version of their Iphone 6 in 2016. Apple launched a series of polycarbonate pastel-colored phones in 2014 (5C) and were received with a great deal of controversy. Apple released the Iphone 5C (C = color) and the Iphone 5S at the same time. The media was extremely critical of the toy-like colors but these phones appealed to a new consumer group of users; those who switched over from other brands and platforms of smartphones. The old Iphone consumers skipped the highly colored phones and purchased the Iphone 5S which boasted higher functionality. The initial sales of the 5S were three times higher than the initial sales of the 5C. This might indicate an interesting strategy where color was used to attract new customers and technology was used to retain old customers. It could also be presumed that the old customers realized that the 5C colored phones would probably need to be covered with a protective case since most smartphones are easily damaged with daily use and the bright colors would not be visible.

Furniture

The furniture market tends to follow function over esthetics so furniture that appears to be more in trend tends to use classic design strategies. However, in recent years, there has been a trend to use natural materials in their natural forms rather than their highly refined counterparts of the past. This is particularly true of Iain Howlett, a British furniture designer who combines wood, bark and bronze metals into very natural but beautiful tables, chairs and lamps. The use of extruded, pulped bark, covering large surfaces of the bare wood creates a rustic element that is juxtaposed with the bronze metal accents that elevates the pieces into the realm of high-end furniture. Howlett is using these materials to bring the outdoors indoors and the use of polished metals harmonizes the furniture with the rest of its interior surroundings.

Furniture fairs play a large role in design weeks sponsored by large cities throughout the world. These week-long design shows feature competitions from students and manufacturers as well as raw material and color suppliers who help furniture designers shape their future products. The color and technology wheels turn very slowly when compared to other industries. This is due, in part, to the long life cycle of most furniture. In recent years, companies like Ikea have sought to change that cycle by introducing less expensive furniture that is replaced every few years. This allows their products to be more on trend than traditional furniture manufacturers whose styles and colors can endure for decades.

Toys

Traditional toys made of wood or metal tend to use bright primary colors. However, more plastic and electronic toys use metallic, fluorescent and special effect colors. This may be due, in part, to the age groups to which these products appeal. Traditional wood and metal, sometimes plastic simple toys are mostly designed for young children under the age of 8. Programmable toys that are radio controlled or multifunctional are mostly designed for older children and are usually manufactured with more sophisticated color schemes.

Lego is an exception to this rule. The basic ABS polymer plastic brick is a very simple building tool but appeals not only to children but to their parents who also played with Legos as children. Basic Lego building sets are still sold with the bright primary colors but the more complex models of things like pirate ships, iconic buildings and motorcycles that are designed with realistic colors that resemble their actual structures in the adult world. Lego has many play sets that are developed just for girls and are designed with very feminine colors schemes of pinks, purples and glitters. Whereas model for the same ages of boys tends to be very realistic shades of bright yellow and black construction equipment or industrial greens, blue and oranges shipping containers. Lego also produces programmable robots called Mindstorms for teens and young adults. These robots are used for competitions between different schools and clubs who program their robots to complete a certain number of specified tasks in a given period of time. Most of these bricks are black and white and little to no attention is currently paid to the color of the robot. Commercial licensing of cartoon characters and movies also greatly influences the color trends and selections that are used in toys.

Consumer Products

Consumer products are broad category of commodity products that include soft goods, durable and disposable products. Soft goods are further defined as products that are made from textiles or flexible plastics. Durable goods are sometimes referred to as heavy goods and are generally defined as products that last longer than 3 years and include such things as washing machines and furniture. Disposable products are those items that are used daily or frequently by customers and last less than 3 years. These are also known as fast-moving consumer products some examples of these products include disposable manual razors, tooth brushes, cosmetics, toiletries, fragrances and their associated packaging. Color and material selections are done in a similar manner to other industries where the marketing group decides on a strategy that best suits new or existing products. The marketing group collaborates with the design group to develop a design brief. The design group either works directly with suppliers to specify colors and materials or they might collaborate with the engineering group to work with suppliers for specification. In some cases, the design group may hand off the project to engineering entirely, and then the engineers work with the suppliers for specifying colors and materials. Once the suppliers are called in and ask to submit sample proposals, they may report only to engineers or the design group or both. This is a longer part of the development process but generally not as long as the testing process where, once the materials and colors have been chosen, the engineers will set up molding trials and other types of stability testing to determine if the specified materials will work for the product design. At this point, the design group needs to sign off on the materials that have passed the testing phase to ensure that the original vision of the design is still a part of the project. If not, a different supplier or types of materials and colors might be re-selected. Once engineering and design signs off on the project, it usually goes back to marketing for final approval that will compare the prototypes to the original marketing strategy to ensure it captures the original intent of the new product or refresh project.

Architectural

Architectural colors and design choices are somewhat different than other industries. This is due, in part, because architects sometimes function as both engineers and designers. Usually, architects will specify products that have already been tested for their type of application by the manufacturer. As an example, an architect that wants to use composite lumber for a decking on spa project will probably specify products from a manufacturer who has already completed extensive testing on the product. In the case of composite or synthetic lumber, the manufacturer has already completed a battery of tests including outdoor weathering, delamination, moisture resistance, fastener durability, swell, warping, etc. At this point is up to the architect to determine if the ratings meet the needs of this project. The architect will also need to determine if the materials meet his esthetic needs of the project. The highest rated composite lumber might not be available in the woodgrains or colors that are needed for that particular project so another grade of manufacturer might be needed to meet all of the specifications of the project. Architects sometimes work with interior designers who provide insight on the colors and styles of furnishings of the project. However, most of the design decisions are typically made by the property owner who might be intimately involved in choosing every color and every material for every segment of the project or the owner might be very less involved, participating in the initial construction brief and not participate throughout the rest of the process. The difference between the architect of a building and the engineer of a manufactured product is that a product manufacturer uses raw materials to create a new product whereas an architect uses finished products to build a larger composite (building).

Further Reading

http://www.cnn.com/2013/10/17/tech/mobile/iphone-5c-flop/index.html
http://www.gottabemobile.com/2013/12/03/iphone-5c-success/
http://fashionista.com/2013/06/top-fashion-schools-2013
http://www.huffingtonpost.com/mary-hall/missoni-for-target-announced_b_857930.html
http://www.luxuo.com/luxury-phones/vertu-for-bentley-smartphone-collaboration.html
http://www.businessdictionary.com/definition/consumer-goods.html
http://www.we-heart.com/2014/10/13/bark-by-iain-howlett/

47 The Antitrend Trend

The idea of tracking and predicting color trends has always been controversial. In the last few years, there has been a backlash against predicting color trends. Personal color selections are very personal. Nominating a color of the year or telling consumers that their color choices are out of fashion can be very polarizing. However, understanding how these colors emerge or what they represent can help chemists, designers and artists determine what color make the most sense for their projects. Part of this analysis is understanding color in historical context. Predicting color trends is similar to predicting the weather. It can change in minutes (like December 7th or September 11th). Another part of this analysis is understanding the technologies that exist to create, measure and apply these colors. Probably, the most important part of this analysis is to understand what is currently happening in various parts of the world and how that will change. In 2014, NY Magazine published an article on Normcore which is simply a trend for people to blend in and look like everyone else. The idea is that all of the emphasis on new trends and individuality is not important to certain segments of the population. This antitrend, unfortunately, is a trend in itself.

Further Reading

http://nymag.com/thecut/2014/02/normcore-fashion-trend.html

Color Trends and Selection for Product Design.
DOI: http://dx.doi.org/10.1016/B978-0-323-39395-9.00047-5

Summary and Conclusions

Choosing the correct color for a product is essential for its commercial success. Color trending can be a good place to start for correct color choices. If a strong color has been in trend for the past few years, a designer might not want to use it for a new product or a product refresh since it can prematurely age a product. The same is true for retro branding where the designer might want to signal to the consumer that the product has, as an example, a 1950s or 1960s feel so it is important that a designer knows what colors were dominant in that time period. Also, even if one is designing a retro product for a certain era, it is important to use colors that were used for those types of products, as an example, 1960s fashion colors were quite bright, almost fluorescent but automotive colors were not nearly as bright. Color trend predictions do not follow a linear path. Global events, natural disasters and new innovations can cause rapid and unexpected changes in color trends. Also color trends for different types of products tend to cycle at different rates. Generally, more disposable or seasonal products tend to cycle more rapidly but not always. Currently, artistic and culture influences tend to be at the forefront of color trends, followed by fashion, beauty and more industrial products. Commodity products tend to be at the back of these trends but that can also shift. An example is in the 1980s when automotive color trends closely mirrored fashion trends but began to lag in the 1990s when financial hardship hit the automotive industry and the manufacturers became more risk averse and offered cars in more limited color pallets.

Understanding the context for colors is also critical. It is important to look at how other products in a brand will be impacted or enhanced by the color chosen by the specifier.

Using color measurement devices are critical for certain types of quality control but should always be checked by eye. In some instances, color measurement devices should never be used, as in the case of products that contain color or effect pigments that tend to the scatter the light and create anomalous dispersions of color. In other instances, optical measurements may negate color differences that are erroneously perceived by the human eye. This is possible when the surface finish or the texture of a product changes and it appears to have shifted the color by visual inspection whereas a colorimeter will ignore the surface finish and only measure the absorption of the colorant and both the standard and the product with the altered surface will yield the same readings, pointing to a difference in surface finish rather than color shift.

It is also important to remember that not all colors can be matched across application methods, as an example, high-gloss metallic finishes are very common in paint or liquid coatings but achieving the same type of effect in an injection molded or extruded plastic might not be possible.

It is should be noted that technical people tend to talk about color in a very different manner than creative design or marketing professionals. Generally speaking, engineers or technical people tend to look and talk about colors in terms of numbers and use more rational definitions for a color whereas more creative people tend to talk about colors in more esoteric terms that relate to feelings or cultural ideas. Be careful of specifiers that only speak in terms of what they do not want the color to be. This is, of course, a generalization and will vary with individuals. This difference in color language can lead to miscommunication and a loss of resources in a developing project. It is also to ask the right questions when starting a new project with a design company or brandowner.

It is important to have a tangible color standard or series of colors that the specifier is hoping to achieve. A physical color swatch is always best if possible, if not, at least a design brief that provides some sort of context for the justification of a color. Even if the specifier only provides a Pantone number, then the supplier should obtain a physical color chip of the desired color if possible. Using digital renderings should be avoided if at all possible since there are massive differences in color shades from different printing devices and computer screens. Also color standards can shift or fade over time depending upon storage temperatures, humidity and lighting conditions and exposure.

Color variation is also different for certain types of colorants. Pigments and metal or salt-based colorants tend to be more stable than dyes. As a rule of thumb, natural, organic dyes tend to be more expensive and less stable than their inorganic synthetic counterparts. Brightly colored colorants tend to be less stable than darker, natural earth tones or metallics colorants.

It is important to qualify the customer as well as the supplier to gage the opportunity at the onset of a project. Just because a brandowner or designer asks for a meeting or requests color submissions does not guarantee resulting business and can be a drain on the color supplier that can draw resources from more lucrative projects that might be a better fit for that supplier.

Certain colors or combinations of colors can set off red flags or send signals to consumers that can change in different cultures. As an example, purple shades can signal death and mourning in Western cultures whereas the same response can be evoked by white shades in Eastern cultures. A similar phenomenon exists for colors that can signal masculinity or femininity that can vary in different cultures.

It is also important to understand that changing a color in an existing product might not be as simple as selecting a new color. As an example, one must take care to think about molding flaws that might occur when adding a pearlescent or metallic effect to an existing color that did not exist in a previous color campaign. Also switching from a dark pigment to a transparent dye can impact the opacity in a finished product. In some instances, dyes cannot molded into certain types of plastics such as polyolefins (PP, HDPE, LDPE, etc.). Also not all colorants are heat stable and will fade, shift or decompose in high-heat applications such as molding or baking. Colors can also migrate or bloom to the surface of a plastic or paint surface when other additives such as plasticizers are present. The chemistry of certain colorants or effect pigments can also change how different layers of a product adhere to each other or using too much of a colorant or an effect pigment can also change the physical properties of the finished product. In some instances, it could make the product softer or more brittle depending on the interaction of the colorants with the substrate. Of course, this is also true for other variations in the design process including modifying wall thicknesses, label changes, switching polymers, reducing, eliminating or adding other additives, etc.

Special attention must also be paid to regulatory requirements since not all colorants can be used for all types of products. Although some colors might be quite stable and suitable for one product, they might be toxic or reactive in another. Also, colorants that might be suitable for one type of product in one region might be unacceptable in another region.

Color selection should occur early in the process so the correct colors and effects can be achieved in the final product. Color selection should be integrated into the initial design and the messaging that the product is hoping to convey to the consumer.

Using the correct color for the environment (indoor/outdoor) is also important. Caution should be exercised to consider the type of lighting used in indoor applications as well as their placement on a shelf or if lighting changes throughout the day, near a window or a door. Understanding how these colors will complement or contrast other products in a series, or how they will be perceived when compared to competitive products is also essential.

All of these components influence color selection and can help to sell the story of future products. Every new product that comes into the market has a story behind it and that story usually consists of why the product was developed, for whom the product was developed, and also the story of the developer and designer as well as their motivations for creating this product.

Additional References

Albers, J., 2013. Interaction of Color: 50th Anniversary Edition. Yale University Press, Haven, CT, ISBN-10: 0300179359. ISBN-13: 978-0300179354.

Abramov, I., Gordon, J., Feldman, O., Chavarga., A., 2012. Sex and vision II: color appearance of monochromatic lights. Biol. Sex Differ. 3, 21, Sept. 4 (open access; highly accessed). (Books and Publications: Peer Reviewed Article).

Birren, F., 1961. Color Psychology and Color Therapy. University Books, Inc, New Hyde Park, NY.

Gettens, R.J., Stout, G.L., 1966. Painting Materials: A Short Encyclopedia. Dover Publications, New York, Series: Dover Art Instruction and Reference Books.

Index

Kolibri 360 GmbH
Ferdinand-Jühlke-Straße 7
99095 Erfurt, Deutschland
produktsicherheit@kolibri360.de